The facts, myths and legends
of the Lost Dutchman Mine
and the Superstition Mountains

by
Robert Sikorsky

•

Golden West Publishers

Front cover photo . . . the Superstitions

Back cover photo . . . Robert Sikorsky at Weaver's Needle

Front and back cover artwork by Bruce Robert Fischer

Maps by Rogga Sikorsky

Photos by the author

Library of Congress Cataloging in Publication Data

Sikorsky, Robert.
   Fools' gold.

   Bibliography: p.
   Includes index.
   1. Superstition Mountains (Ariz.)--History.
2. Gold mines and mining--Arizona--Superstition
Mountains--Folklore. 3. Waltz, Jacob. 4. Sikorsky,
Robert. I. Title.
F817.S9S54   1983   979.1'75   83-11626
ISBN 0-914846-15-9

Printed in the United States of America

Copyright © 1983 by Robert Sikorsky. All rights reserved. This book, or any portion thereof, may not be reproduced in any form, except for review purposes, without the written permission of the publisher.

Golden West Publishers
4113 N. Longview Ave.
Phoenix, AZ 85014, USA

DEDICATION

For Rog,
who has known this book almost as long as I have.

Acknowledgements

I would like to acknowledge the assistance and/or resources of the following persons and institutions who contributed in some way to the building of this book:

Lynn Bailey, for his early editing and whose recent suggestion triggered the eventual publication of this book.

The Arizona Historical Society, Tucson, and its competent staff, especially **Margaret Bret Harte,** head librarian, who has pulled many files for me through the years.

Dr. Willard C. Lacy, former head of the Department of Mining and Geological Engineering at the University of Arizona, who did the rock analysis for Chapter 10.

Donald A. Van Driel, District Ranger, Mesa Ranger District, for his help and comments on Chapter 16.

The staff of the **Mesa Museum,** Mesa, Arizona, especially **Angie Rosales,** for the many courtesies extended.

Rev. Charles W. Polzer, S. J., ethnohistorian at the Southwestern Mission Research Center of the Arizona State Museum in Tucson, for his incisive comments on the "Peralta" Stones.

Daniel J. W. Huntington, early Arizona Pioneer, for his boyhood recollections of Jacob Waltz.

The University of Arizona Library and the **Tucson Public Library** staffs.

The personnel at the **Arizona State Library,** Archives Division (DLAPR), Phoenix, and the **Maricopa County, Yavapai County and Pinal County Recorder's offices.**

The Wells Fargo Bank, History Room, San Francisco; **University of Heidelberg,** Records Division; **General Services Administration,** National Archives, Federal Records Centers at Denver, San Francisco and Washington, D. C.; **United States Mint,** Office of the Director, Washington, D. C.; **A. L. Flagg Foundation,** Phoenix.

The many currently operating newspapers from which material was gleaned, especially **The Arizona Daily Star, The Arizona Republic, The Phoenix Gazette** and **The Apache Sentinel;** the numerous newspapers of yesteryear that helped make the journey into the past possible.

The U. S. Bureau of Mines; The U. S. Geological Survey; The Arizona Bureau of Mines; The Arizona Department of Mineral Resources.

And lastly,**Maria, Louie** and **Raymond,** wherever they are.

Contents

Vignettes
1. "... someone who could tell one rock from another..." 7
2. "Hi, I'm Maria Jones." 13
3. "Fortune is smiling..." 18
4. "I wanted action." 24
5. "... twirling... like a spider... on the end of a thread." 31
6. "... not me, Maria." 41
7. "... slipped and fell some thousand odd feet..." 46
8. "... the man in Scottsdale was one of her 'angels'..." 53
9. "... shooting it out." 66
10. "... using dynamite... on the Needle" 72
11. "... one night he would never forget." 76
12. "... people on the Needle..." 81
13. "... the most forbidden country..." 94
14. "Somebody... was... following me." 108
15. "... someone was shooting into our camp." 118
    Epilogue 127

Chapters
1. In Coronado's Wake 9
2. A Brief Journey 15
3. Yesterday is Forever 19
4. Lifting the Veil of Time 24
5. Jacob Waltz 36
6. The Dutchman's Darkest Day 42
7. The "Superstitious" Indians 47
8. The Eternal Search 55
9. Earthquake!! 69
10. A Taste of Geology 73
11. "Lost" Mines 78
12. Death is My Nickname 83
13. The Most "Found" Lost Mine 95
14. The "Peralta" Stone Maps 110
15. What Happened to All the Gold? 121
16. A Look at Tomorrow 124
    Bibliography 128
    Index 141

Maps
   Superstition Trails... Inside front cover
   Superstition Wilderness... 6
   Earthquake Zone... 68

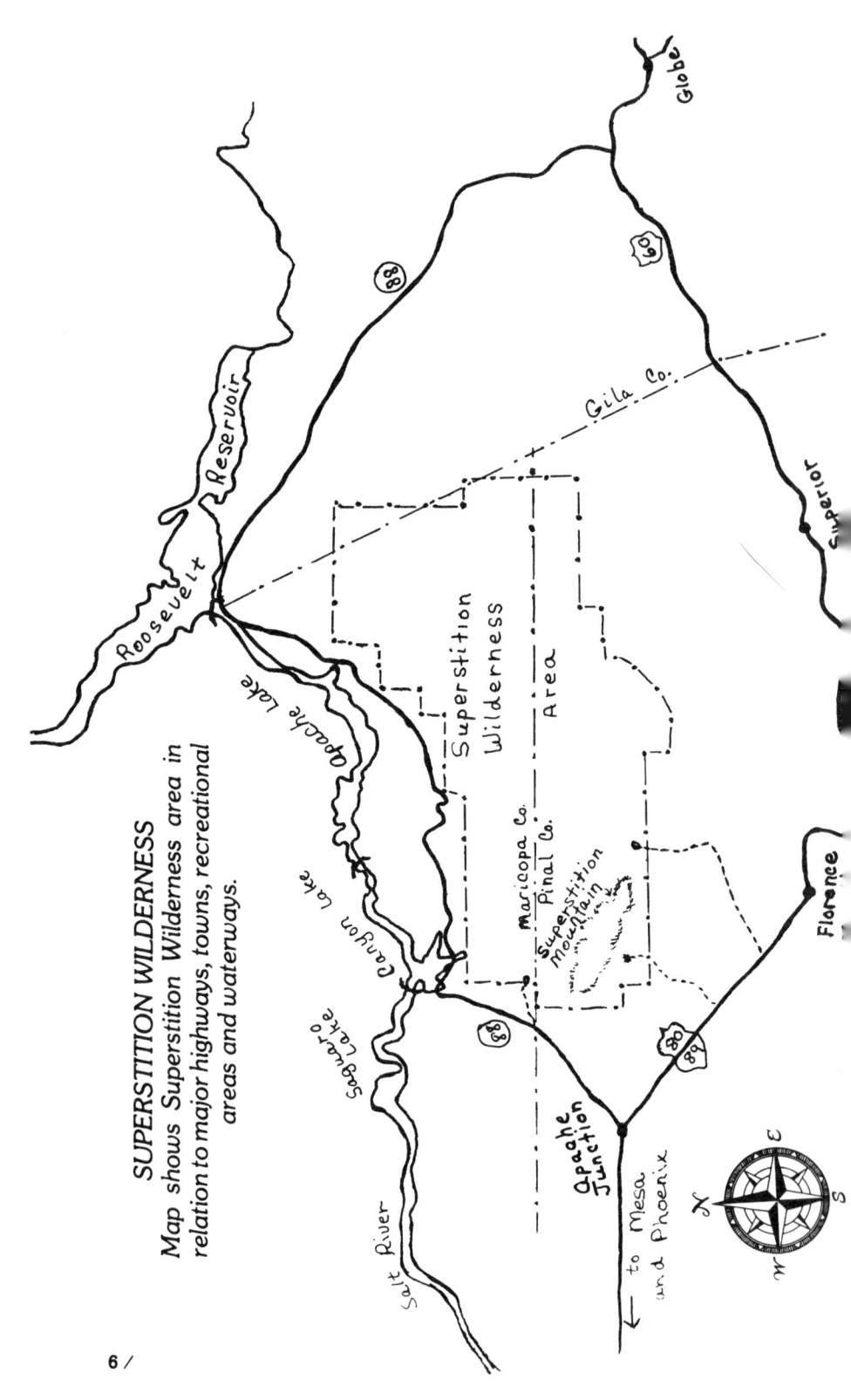

SUPERSTITION WILDERNESS
Map shows Superstition Wilderness area in relation to major highways, towns, recreational areas and waterways.

vignette 1
"...someone who could tell one rock from another"

Early in 1959, not long after I had arrived in Arizona, I answered a "Help Wanted" ad in a Phoenix newspaper. The ad was looking for a geologist or some person with geological experience. I had a passing knowledge of geology, especially mineralogy, and had come to Arizona to study it at one of the universities. Something prodded me to answer the ad. I called the number and the man who answered gave me an address and asked me to come the next morning for an interview.

The following day found me at a very small house on Adams Street in the central Phoenix area. There I met a Mexican man who introduced himself only as Raymond. He was the interviewer. He asked me a number of questions, one of which was "What is the piece of metal I am holding?"

I took it from him, studied it for a moment and told him it was pyrite or "fool's gold" as it is commonly called. He nodded but didn't say anything, his eyes riveted on the piece of metal as I handed it back to him. I thought for a moment that there was a brief look of disillusionment on his face. Surely he didn't think this was gold? After turning the metal over and over in his hands for what seemed like an eternity, he placed it in a large ashtray full of similar smaller pieces. He then turned toward me and asked if I had ever heard of the Lost Dutchman Mine.

The Lost Dutchman! Of course I had heard of it. It was pretty hard to be in Arizona and not hear of it. In the short time I had been here, I had listened to many tales about the lost gold mine in the Superstition Mountains, most of the stories coming from some Mexican friends of mine who lived in Superior, a small copper mining town near the eastern extremity of the Superstitions. I was staying with them in Tempe, where they were attending the university. Some of their relatives had actually looked for the mine. They all believed that there was a lost mine somewhere in the Superstitions and that it was an extremely rich one. Their stories about the mine fascinated me, and I had even gone out and bought some booklets about it. This sharpened my curiosity even more.

"Yes, I've heard of the Lost Dutchman," I told Raymond.

"How would you like to get paid helping someone look for the mine?" he asked.

"You want me to help someone look for the mine? I've never even been in the Superstitions."

"It doesn't matter," he replied. "We are just looking for someone who can tell one rock from another. Our crew knows the mountains like the back of their hands. We don't need a guide."

Before I could mutter a stifled "yes," we were shaking hands and sealing my commitment. No papers to sign in this dusty rundown central Phoenix "office." A handshake and a "yes" were all that was needed to buy a ticket to the Superstitions.

Raymond told me I would not be working for him, but for another person whom I would meet the next day, a person he said had reliable information that pinpointed the exact location of the Lost Dutchman Mine!

Photo taken by Maria Jones

ROBERT SIKORSKY
The author as a young college student is given opportunity to study geology with Superstition gold seekers.

1

> *Granted that they did not find the riches of which they had been told, they found a place in which to search for them.*
> **Castañeda** - *Chronicler of Coronado's Expedition*

In Coronado's Wake

By 1500 Spain had been caught up in a state of frenzy and excitement. She had become a ruler of half the world, and her armadas were still probing unknown and mystic regions. It was as if someone said, "Here, take the world, it is yours."

Allegedly bursting with treasures, the new world waited. In Spain, stories had long been told of seven great and gilded cities that existed somewhere beyond the horizon—the legendary Seven Cities of Cibola. So rich and beautiful they were supposed to be that bowls and dishes were made of pure gold and walls of houses were plated with it. But the seven golden cities were somewhere beyond, *más allá*—in an unknown region.

By 1531 the Spaniards were searching the continents of the New World with great zeal. In Peru, Pizarro had already wrested from the mighty Incas an entire empire which had taken centuries to build. In Mexico, the proud Aztec nation was being brought to its knees by Hernando Cortez, and the city known as *Tenochtitlán* became Mexico City, and was destined to become the hub of the New World and center for an expanding arc of conquest.

While the *conquistadores* roamed the western side of the continent, eastern shores also experienced the same persistent activity. Florida was discovered and new vistas opened there. But these foreign lands were not opened without struggle and many explorers gave their lives, as if in payment for intruding on the new territories. On one occasion, a ship exploring the Gulf Coast of Florida was caught in a vicious storm. After days of being tossed around like a toy, it was finally wrecked on a distant shore, spewing contents helterskelter and leaving but four survivors: Nuñez Cabeza de Vaca, Andres Dorantes, Castillo de Maldonado and the black Moor slave, Esteban.

Building a raft from the wrecked ship's timbers, the wretched foursome set out with benign hopes that by sailing west they eventually would reach Spain. Instead, after many hazardous days on the small raft, they reached the Gulf Coast of what is now the state of Texas. Still with hopes of reaching Spain, they set out on foot in a westerly direction. To say the least, their course was a wandering zig-zag on which they became acquainted with many new and strange peoples—the Indians of the southwest. They pushed onward, finally reaching the Rio Grande river and then crossing into Northern New Spain. Their imaginations had been set afire by reports that many rich towns existed to the north, golden towns with strong proud inhabitants, so rich that even the streets were paved with gold. The stories that had been told for so long in Spain were really true. The Seven Cities of Cibola were real—and they lay but a short journey to the north. *Más allá.*

After six long and incredible years of wandering, the errant group was found by a party of explorers sent out from Mexico City to investigate the lands of the north. Their meeting was a memorable one, and after much ceremony and celebration, plans were made for a hasty return to Mexico City and a report to the Viceroy. Enthusiasm was high, for the reality of the golden cities had now been confirmed and it would be a short time before they yielded their riches.

In Mexico City, the tales by de Vaca and his party captivated Spanish officialdom. The Seven Cities of Cibola—although as yet not seen by Spanish eyes—became as real and vivid as the sun above. Upon receiving the amazing news, Antonio de Mendoza, Viceroy of New Spain, realized that here was an opportunity to bring glory to himself and enrich the Spanish crown. An expedition was immediately organized; and in 1539 an advance guard set out to map the best route northward. It consisted of Fray Marcos de Niza as leader; Esteban, whose incredible journey was still vivid in his memory; and a group of Indians who were to act as porters and messengers. The company trekked steadily north, Fray Marcos diligently recording all that was seen and encountered along the way. At one point in the journey he became the first white man ever to set foot in the present state of Arizona.

As the party pushed further north, Esteban became more and more enthusiastic, for he knew that the golden cities were near. Fray Marcos, not wanting to contain the Moor's excitement, sent him ahead with several of the Indians; his mission was to verify the existence of Cibola. If the news was bad, he would send back a small cross; if good, a large cross would be the sign. Fray Marcos himself stayed with the main party and enjoyed a more leisurely pace.

A seemingly short time later, the signal from Esteban was received; an Indian had returned carrying a cross as big as a man! The Seven Cities of Cibola had been found! Fray Marcos was exuberant—it was the news everyone had wanted to hear. The report, however, dampened, when a few days later one of the Indian guides returned with the sad tale that Esteban had been killed by the inhabitants of the golden cities. Upon hearing this, de Niza hesitated, and after little deliberation as to what action to take, decided to retrace his steps to

Mexico City and not risk contact with the inhabitants of Cibola. After all, he had received the good news from Esteban and that was most important. Once in Mexico City he could report to the Viceroy that the Seven Cities of Cibola, which he himself had never seen, were real.

Mexico City received the news with disbelief. The Viceroy was beside himself. A grand expedition would be organized; hundreds of mules, horses, and men, loads of equipment. A *carreta grande!* Fray Marcos, his recent journey fresh in his mind, was chosen as guide and other missionaries were to accompany him, for many pagan peoples lived along the way. As leader of the expedition, Mendoza chose Francisco Vasquez de Coronado, a strong and persevering man he trusted.

Early in 1540, the party set out and after many months of tiring journey, they too crossed into Arizona, where not long before Fray Marcos de Niza had been. The trail, bearing ever northward, continued up the lush and fertile San Pedro Valley. But, by now the grumblings of the men became louder as food and tempers grew short. Coronado himself became irritated and impatient: the supplies, promised to reach him by ship via the Gulf of California, never arrived. His party was much further inland than he and the Viceroy had anticipated, and the coastline, along with the supply ship, lay far to the west.

Finally, the months of hardship and weariness drew to a close. The Cibola were near at hand! Elation and anticipation were high as the expedition pushed its way over the crest of the final ridge. At last, after all the years of speculation and hope, there they stood, the fabulous Seven Cities of Cibola, the sun beating brightly on the mud walls of the small huts and glistening on the near-naked bodies of the inhabitants!! The seven golden cities were nothing more than the wretched mud pueblos of the Zuni Indians, and the only gold to be found was in the rays of the burning sun! The depression that engulfed Coronado and his men was boundless. And Fray Marcos, who earlier had reported the gold cities to be real, was castigated to such an extent that, as of this point, nothing more is said of him in the historical records.

The theme would be a familiar one for centuries to come. It was destined to have many variations, but results would always be the same. The curse of Coronado had been placed, and from that time on those who dared hunt for treasures would have to contend with it. To be sure, it was this very notion that prompted J. Frank Dobie to call those who followed "Coronado's Children."

However, disenchantment was soon turned into hope for Coronado, when from the Hopi Indians he learned of great peoples and cities to the east. There, not here, lay the *Gran Quivira; más allá, más allá.* Coronado pushed onward, revitalized by the winter's rest. He journeyed far east, into present-day Kansas, again only to find that the *Gran Quivira* was just a handful of Indian huts. Coronado could not escape his own curse. Downtrodden, disenchanted and beaten, he turned back. He had news to present to the Viceroy—the golden cities and the golden dream did not exist. Upon return to Mexico City, Coronado's report was greeted with disbelief. Men refused not to believe the tale which had taken centuries to build. The cities just had to exist. Perhaps Coronado had searched the wrong region; a little more to the north; maybe to the west might lay the realization to their dreams. They had to be somewhere.

Men cling strongly to what they have earned, and indeed the Spanish had the right to believe in a dream that so many helped to build. It would take more than a mere expedition to crush their hopes. Castañeda, the chronicler of Coronado's expedition and astute observer of the ways of man, put it well when he said: "Granted that they did not find the riches of which they had been told, they found a place in which to search for them." Indeed a place to search! A place so vast and unexplored that the ensuing centuries would give birth to more stories of lost and hidden riches than all the other parts of the world combined. Coronado had done his job well.

HISTORY
What bold plans or tales are told in Superstitions petroglyphs?

"Hi, I'm Maria Jones."

vignette 2

I didn't sleep that night—an experience I would repeat on my last day of employment also—thinking about the strange interview. This was not life as I had been used to in my native Pennsylvania.

The next morning I drove back to the Adams Street house. As I pulled into the back yard, I could see Raymond sitting on the small porch talking with a little mustachioed man dressed in khaki. I wondered if he was my employer. The two came down to meet me and Raymond introduced the little man. He was "Louie" and he wasn't the one I would be working for. As we exchanged greetings, the back screen door swung open and a large, sunglassed black woman appeared.

"You must be Bob," she drawled. "Hi, I'm Maria Jones."

Maria Jones. The same Maria Jones who, unbeknownst to me at the time, was already well-known to followers of Superstition lore and somewhat of a legend herself. "Celeste" Maria Jones, the second black woman to look for the Lost Dutchman, an alleged professional singer, a minor cult figure, a person who fascinated the media, had already been in the Superstitions for quite some time, searching for the mine. I thought it strange that a woman should be the head of this operation and I wondered if she actually went into the mountains herself. As I found out later, she was quite at home there. Unknowingly, I had hooked up with one of the most bizarre groups ever to set foot in the Superstitions, and I would, as I found out later, be doing more than just looking at rocks. (In almost all media accounts "Maria" is listed as "Marie," but to me and the others in her camp she was "Maria.")

Maria told me that her grandfather was related to someone who knew the Dutchman, Jacob Waltz, personally. I later wondered if this could have been Julia Thomas, who was the first black woman to look for the lost mine not long after Waltz had died. Sitting on the porch, I listened to her as she talked about herself and the Lost Dutchman Mine. She had studied music at Juilliard, she had sung to big audiences, she had been around. She showed me some pictures. Singing in a nightclub? A concert hall? At the microphone in front of a large orchestra; another standing by a fancy car, dressed to the hilt. She was quite a bit younger in the pictures, and thinner too, but there was no doubt it was her. I wondered at this person I had just met. She showed me a scrapbook—clippings and pictures of her singing, one story praising the clarity and purity of her voice. What then was she doing looking for a lost mine in the Superstitions? I couldn't make the connection. And what was true and what wasn't? It was hard to tell, and as I would find out later, it would become harder.

My first reaction after meeting this crew was to turn and run, get away from it before I became involved. The stories I had just heard, the people themselves, were all too unlikely for me to digest, too much for this Arizona newcomer. But the whole affair fascinated me. I had to see what lay beyond this initial encounter; perhaps there was some truth to what Maria, Raymond and Louie said, perhaps they did know where the lost mine was located, and, ever since listening to the stories of my friends from Superior I had wanted to go into the Superstitions. I'd never get a better chance!

2

A Brief Journey

Many eons ago, when the earth's crust was still cooling and its features being determined, there erupted in the state of Arizona a gigantic mass of molten lava and jagged rock. A mountain was born that would figure heavily in the lives of generations of men who would live and die near it. From its depths the earth poured forth vast quantities of copper, gold, and silver, and along with other rich metals and minerals, mixed them with the surrounding molten rock and deposited the results in gigantic pockets. It was on or near these vast ore bodies that towns would spring up and people would live, work, and die, giving a great part of their lives to extracting these precious metals. This area would be known as the great Globe, Miami, and Superior mineral belt and from it would spring such famous mines as the Silver King. To the west of these deposits and immediately adjacent to them was another mass of rock. At first, the Indians who lived there called it Crooked Mountain because of its gnarled and twisted shape; then the Spanish came and named it *Sierra de la Espuma* or Mountain of Foam, after the white limestone smear across the top; finally, American settlers named it Superstition Mountain, for reasons as mysterious as the mountain itself. Only some forty odd miles east of present day Phoenix, Arizona's largest city, the Superstitions cloaked themselves with a veil of mystery even the passage of time cannot remove. Here in this wilderness, there didn't seem to be any substantial mineral wealth, as most of it had already been placed to the east, near Globe and Miami. Just a few miles from the western edge of the Superstitions, however, the small mining town of Goldfield caused a brief sensation when much gold was mined from beneath the town's surface. The vast interior of the Superstition Mountains seemed devoid of mineral wealth. It was here though, in an area thought to be very high in geologic improbability, that a search started and continues to this day—a search for the richest gold mine in all the world—the fabulous Lost Dutchman.

The main body of the Superstition Mountains is composed of a rock called dacite, a volcanic combination of quartz, plagioclase feldspars, hornblende and sometimes biotite, that dates back some sixty million years. On the geologic map of Pinal County, in whose confines most of the Superstition Mountains lie, there are indications of basalt and rhyolite formations. Weaver's Needle, an important landmark associated with tales of the Lost Dutchman mine, is marked as a dike or plug, a volcanic uplift that may be likened to cork in a bottle. Near the town of Goldfield, the map notes a few small mines, but other than that there is no professional recognition of mining activity anywhere in or near the

entire Superstition range.

As one travels by car from Apache Junction, a prospering desert community located in the shadow of Superstition's western edge, to Florence Junction, some 25 miles to the east, one drives a modern four-lane highway that follows the southern edge of the mountains. Along this highway, one is compelled to gaze at the magnificent Superstitions, just a few miles off the highway to the north. While traveling this road, if one turns his car north, he will arrive at King's Ranch, a famous resort snuggled near Hieroglyphic Canyon, the top of which marks the Superstition Mountains highest point, 5,057 feet. Arriving at Florence Junction and continuing east for approximately fourteen miles one comes to the small mining community of Superior. Northeast of here are Miami and Globe. This is the mining district where the famed Silver King once thrived. At Globe the road again turns northeast and Route 88, the scenic and aptly-named "Apache Trail" is taken, and soon Tonto National Monument and its remnants of pre-historic Indian dwellings can be seen. The colorful drive continues as the road follows the meandering Salt River. Eventually one is greeted by views of Roosevelt, Apache, and Saguaro Lakes, vital man-made reservoirs that feed precious water to Phoenix and the Valley of the Sun. Traveling this portion of the Apache Trail, a visitor encounters some of the most splendid scenery to be found in all Arizona.

From the lakes, the Apache Trail turns south, back towards Apache Junction, the original starting point of the journey. It is this portion of highway that affords the most magnificent view of the Superstition Mountains.

When the setting sun flirts with the orange-colored face of the mountain, it is easy to see how the name "Superstition" originated.

Along the mountain's western extremity, gigantic spires are thrust from the desert floor, beckoning the sky above. If one looks closely from here, he can see the many figures on top of the great jagged cliffs. These, according to a legend of the Pima Indians, are of people turned to stone. Also discernible is the white limestone streak along the top of the cliffs which the Pimas believe to be the height attained by the legendary flood of their creation myth.

Completing the journey at Apache Junction, the point at which it started, a giant circle has been traveled, and the entire Superstition Mountains have been circumscribed. At most points along this trip, portions of the Superstitions can be seen, their nearness impressing the casual visitor and even life-long residents. If in reasonably good shape, a person can park his car at the end of a dirt road that juts off of the Apache Junction-Florence Junction highway, and in a matter of two or three hours of leisurely hiking, be at the base of the famed Weaver's Needle, the giant rock pinnacle deep in the heart of the Superstition range. Ready access can also be found at the west side of the mountains, where, while traveling along the Apache Trail highway, a party turns off at First Water, near the point where the mighty front face of the Superstitions terminates itself. From here Weaver's Needle is a three to four hour hike.

The relatively easy access to the mountains has undoubtedly been one of the major factors in luring people to hunt for the Lost Dutchman Mine. Supplies can readily be purchased at Apache Junction or Mesa, some fifteen miles west, and ample guide services are available. From there it is just a literal hop, skip, and jump into the land of the legendary Lost Dutchman gold mine. The calendar can be turned back a hundred years in just a few hours walk or horseback ride, and the same trails taken by old Jacob Waltz, the Dutchman, can be traversed.

But, it should be mentioned here that the easy access to the center of this fabled country can be **deadly** deceptive. Once in the mountains and off the main trails, searchers for the lost mine are greeted by some of the most rugged, hazardous, and inhospitable terrain in the world. Steep cliffs, rocky wind-tortured and heat-infested canyons, scarcity of water, rattlesnakes, death; all are there. Caution should be the byword.

It has generally been agreed that two things are vital if a trip to the Superstitions, whether for pleasure or prospecting, is in the offing. Number one—and most important—is to **never go in alone.** Number two, is to **let someone know your plans, your destination,** and **the date and time of your anticipated return.** Too many people have ventured into the Superstitions without first abiding by the above stipulations and were never heard of again. Don't be one of them! Don't underestimate the Superstitions!

vignette 3

"Fortune is smiling..."

That evening Raymond called to say that he would be picking me up early the next morning. I packed some gear in an old Marine knapsack, and, as Raymond had suggested, brought along my Husquavarna army revolver, a recent pawn shop purchase.

Morning found Raymond, Maria, Louie, me and a load of gear crammed into a powder blue 52 Chevy. From Tempe, where they had picked me up, we headed back toward Phoenix and not toward the Superstitions. I felt a little uneasy until Maria said we had one more stop to make before heading for the mountains.

Raymond guided the car to another rundown shack, this time in south Phoenix. As he parked the car, we were greeted by an apparition right out of a gypsy caravan. Old and gnarled, wearing a loose flowing dirty dress, shoeless, with long fingernails and a shawl wrapped around her head, the gypsy came out to meet us.

"Bob, this is Nadine," Maria said.

She nodded and broke into a toothless grin, her leathery hands cupping my face.

"You will be good for this group," she said.

We followed her into the house and took seats around a battered wood table which somehow seemed out of place amidst the clutter on the floor and shelves. As we sat, Nadine the gypsy began to mumble some strange incantations as Maria, Louie and Raymond bowed their heads. I bowed mine. She was muttering in Spanish and I couldn't understand what she was saying. When she finished, everyone looked a bit relieved, as if an important chore had been completed.

"You will have good luck this trip Maria," the gypsy said. "Fortune is smiling on your group."

She said it would only be a matter of time until we were all very rich. Then, like a priest blessing his flock, she stood up, opened her arms and raised her head and uttered a few more words in Spanish. With that, the ceremony was complete and we headed toward the car. Maria, Louie and Raymond seemed somehow fortified by the visit to this strange woman. What had I gotten myself into, I thought, as the car pulled out and headed toward the Superstitions.

3

Yesterday is Forever

The Spanish were really the ones to blame. Prior to the advent of the *conquistadores,* the Indians of Arizona and the West had very little knowledge of and even less use for the metal of the sun, called *oro.* The desert Indians were much concerned with the daily struggle for existence, and had little time to devote to mining. Objects of gold and silver, when they did turn up, were considered oddities and attributed to a monentary whim of the owner. A few pieces trickled in via the routes of trade, but generally, gold and silver had no intrinsic value to the Indians, and to collect something worthless was foolhardy.

It was not until the appearance of the Spaniards in the Southwest that the Indians began to realize that the soft gold metal was something unique. The zeal with which the Spanish combed the area looking for gold and silver knew no limits—literally, no stone was left unturned.

In many cases the Indian, intimately familiar with the landscape, would prove valuable, guiding the explorers to places where the valuable metals outcropped to the surface. Many of these sites were eventually turned into very profitable mines. The Indian slowly began to accommodate and absorb the ways of their "guests," and although still fiercely loyal to the "old ways," realized there was value in knowing how the white man operated and what he held most important. The peaceful valley dwellers were most coveted by the Spaniards. Gold and silver was wrought from their land and new religious beliefs were foisted upon them. Precious metals and conversions were what the Spanish wanted and the placid natives were essential to their secular and religious plans. In time, many of the ancestors of these same Indians would be central figures in the many lost mine and treasure tales that would come out of this land they once called their own.

The Indians, now becoming used to the presence of the white man and his strange ways, were slowly being converted to Christianity. They were on their way to becoming "civilized" and the mineral-bearing provinces of northern Mexico, southern Arizona, and New Mexico were rapidly being colonized. Missionaries and miners worked together and the Indian was taught mining and the Christian faith. The zeal of the Jesuits had no limit—conversions for the Church and precious metals for Spain. The spiritual and mineral resources were heavily tapped.

At times, the Indians rebelled, but superior forces of the Spaniards always brought them back into line. But one Indian tribe never gave up, and their presence would be felt during the entire Spanish occupation. They were called "Apache," a word used by the Spanish to denote an enemy. A better name could not have been bestowed, for "Apache" would ring out through the centuries and would signify the fiercest enemies.

Thus the scene was set. Northern New Spain, a vast and desolate land, slowly gave of its abundance to the persistent Spaniards. Coronado, in all his disillusionment, never dreamed of what was to come. The seven cities of gold were never found, but the amount of wealth extracted from this "place in which to search for them" was enough to build seven cities many times over.

While the Spanish were busily engaged on the western side of the continent, the East was giving birth to a new nation. It would not be long before this country, eager to expand and bursting with energy and adventure, pushed its borders to the edge of the Spanish empire. Here the two met; one old, tiring and very rich; the other new, vibrant and very poor. The rest is history. The United States, not content to be static, extended its borders from Atlantic to Pacific and from Canada to Mexico. Spain had yielded; she was content with money and Mexico.

Besides lingering Spanish activity, there was little ado in the West up to the early 1800's. The few who had come journeyed by ship, around the Cape, and settled in now famous locations along the California coast. Some attempts were made to cross overland, but these journeys were accomplished by the brave and adventurous, and were few and far between. It was not until word of gold discoveries in California filtered back to the East, that any concentrated effort was made, via overland, to reach the West. Almost overnight, California became the promised land. "Go west, young man" was the cry. And they went.

An often-ignored fact about Arizona is that during the early and mid-1800's it was not a great center of mining. Spanish and subsequent Mexican activities had almost ground to a halt; and Arizona was but another dry and rugged expanse that had to be crossed in order to reach California.

The name "Arizona," probably derived from the Papago and Pima Indian dialects, means "Little Spring." "Ari" means small and "Zonac" spring. But in reality, all the land of the Little Spring had to offer was heat and dust, sweat and wilderness, and even as late as 1862, when Arizona was pushing to become a territory, its population was estimated at a mere 2,401, and most of these people were described as "miners seeking adventure and wealth." Many in Arizona were those too weak or too broke to push on to California. Hindered by their inability to reach the alluring gold fields to the west, they were forced to eke out an existence in the barren country: prospecting, lumbering, merchanting and promoting. Nevertheless, they laid the weak and wobbly foundation for what would someday become the fastest-growing state in the union.

By mid-19th century many stories of Spanish mining, conquistador treasures, Aztec gold, and lost mines had become commonplace. The Aztecs, unlike their Indian neighbors to the north, had had many uses for gold. Indeed, they worshipped it! History leaves little doubt that the Spaniards confiscated large amounts of Aztec treasure and shipped it back to Spain. But was *all* of the wealth accounted for? Did the Aztecs hide *some* or perhaps even the *greater part* of their treasure? Was Montezuma cunning enough to have tricked the Spanish into believing that they had all of the Aztec gold, when in reality they had only a small part? Questions and responses, stories, deeply nourished by Spanish and Indian traditions, passed swiftly from mouth to mouth. Tidbits of information trickled about, tantalizing the imaginations of the hardy men. Everyone knew that the Aztecs had hidden their hoard from the Spanish—but where? Sealed and well-hidden Jesuit silver mines existed near Tucson and Nogales and waited to be found. In northern Mexico, the hidden silver of Tayopa lay waiting. Indians whose ancestors had once worked in the Jesuit mines told stories about the wealth they contained. Some even had vague directions passed on to them by word of mouth from their elders. Many attempts were made to find the mines, but all were frustrated in the end. Nearly 300 years had passed since Coronado marched this land and his children were still searching. Somewhere that dream was hidden, and someone, some time, would find it. It just had to be so! And it was in this environment that the legend of the Lost Dutchman Mine took its first fragile roots.

The greater part of the story of the Lost Dutchman did not materialize until after the death of the Dutchman himself, Jacob Waltz; but in order to better understand the central theme, it is necessary to become acquainted with the introduction. *That the Lost Dutchman Mine is legendary and is not an historical fact* hinders one from uncovering any truths that may lie hidden in its past.

But, what is a legend? Webster defines it as "a story coming down from the past; especially one popularly regarded as historical although not verifiable."

The saga of the Lost Dutchman Mine fits the above definition perfectly. It is a story that cannot be altogether verified; its origins are hidden somewhere in the past. It lies somewhere on the fine line that separates fable from recorded history, and it has drawn freely from both sides, embodying itself with quantities of fact and fiction. The story of the Lost Dutchman Mine can never be verified in all its details; there are too many intangibles, too many contradictions and too many versions. But these are the substance that makes the legend imperceptible to the ravages of time. Thus, the people who wish to believe in the legend can do so freely, for they are just as correct as those who scoff at it.

THE CHANGING FACE OF WEAVER'S NEEDLE

… vignette 4

"I wanted action."

My first days in the Superstition Mountains were ones filled with excitement and wonder. It was a landscape and environment completely foreign and it impressed me tremendously. I was awed by the vast rugged range and its majesty and aloofness. There was something secretive here, something that said these were the Superstitions and not just another mountain range. It was easy to see how they became the spawning ground for untold stories. Secrets seemed to hide in every crag, whispers behind every boulder—taunting, cajoling, daring you to uncover the mystery of the Dutchman's lost gold. It wasn't hard to visualize oneself coming across the mine at any moment.

Raymond stayed in camp but periodically and I learned that he was the liaison between the mountains and Phoenix, the transportation man, ever so important to the operation. Louie must have been born in the mountains, for he could move across the terrain like a water bug across a pool of water. I was impressed with his quickness and grace and his ability to carry a large load on the trail. His stature belied him. Maria impressed me also with the ease with which she ambled over the rocky ground. They knew the mountains well, these two, for they had already been here on and off for a number of years.

I quickly became acclimated to camp life and the terrain around it became familiar. The camp was in Boulder Canyon, just south of the majestic spire called Weaver's Needle. Looking up at the Needle, I could see how it became a central landmark in the Dutchman legend—it was the most visible and most impressive piece of rock in the interior of the Superstitions, its top reaching nearly 5000 feet. Camp life was easy and relaxed those first days and it bothered me that I had heard nothing about looking for the lost gold. I began to get anxious, I wanted to explore more of the Superstitions, I wanted action. Louie said that I would get my chance real soon for tomorrow we were going to do some serious climbing. He was going to show me the place where Maria thought the lost gold was hidden! I would get a real taste of what the Superstitions were all about. He wasn't kidding!

Lifting the Veil of Time

4

The Lost Dutchman legend has undergone many changes through the years. The story has been added to; parts have been altered—it, too, has adapted to the changing times. It has been flexible, withstanding the ravages of nearly a hundred years.

A "lost mine" in 1860 may have been assigned an arbitrary value of $1,000,000; today, the same "mine" may have increased to ten times that. We sometimes play uncanny tricks on ourselves because something lost for that period of time, in reality, has very little chance of ever being found. Nevertheless, as the years pass, more people learn of the lost bonanza and invariably the value will automatically increase.

The passage of time only adds more and varied versions to the original story of the "lost mine," because of the greater number of people familiar with it. It is almost a truism that the more people who know about a legend, the less they know. It's a lot like some of the television panel shows in which someone is told a bit of information and he in turn is instructed to pass it on to another. Perhaps this transfer of information takes place four or five times, and finally, the last person to receive the news repeats it to the original teller. It is amazing to observe how much the original version has been changed— and in such a short period of time.

Imagine what a hundred years and thousands upon thousands of people have done to the Legend of the Lost Dutchman Mine! Stroll through Phoenix or Tucson today and ask someone if he knows anything about the Lost Dutchman Mine. More likely than not he will give you a story that will vary considerably from that of the next person.

Fortunately, the Lost Dutchman Mine has left some traces of itself and the people most intimately associated with it in the written records of the past. In the newspapers of yesteryear are some of the brief accounts of the mine and of Jacob Waltz, the man most commonly associated with it. The reader will now take a journey into the past, back to the time when Jacob Waltz, the Dutchman himself, roamed the Arizona deserts. The trip will be taken via the newspapers of the time and an intimate glimpse will be had of what people then thought of Jacob Waltz, the Lost Dutchman Mine, and the Superstition Mountains.

Put the coffee on, crawl into your favorite chair and relax. The journey is a fascinating one.

The Weekly Arizona Miner, Prescott, February 15, 1878:

Two Germans arrived in Florence on the 6th, from a prospecting tour through Superstition Mountain, where they found an extensive silver ledge, the ore from which assays over three thousand dollars to the ton. 'Old Superstition' is in Maricopa County, consequently the district which will be formed at the new find will add another district to the list in that county. The Germans who made the discovery, found, at the base of the mountain, on their way in, the remains of a white man, who evidently had been killed by the Indians. His body was cut up and mutilated to such a degree that a description or recognition was out of the question. Many are preparing to visit the new silver fields.

In this short article we can sense the excitement of the time. Prescott, the booming mining and lumber capital of the Territory of Arizona, is being swept with a frenzy of mining talk, as new districts are being formed, new discoveries made, and new claims filed. Activity is rampant and the air is electrified

The reference to the two *German* prospectors in the Superstitions is intriguing. Could they have been Jacob Waltz and his partner Jacob Weiser? If so, this would be one of the first references to Waltz that associates him with prospecting in the Superstition Mountains. And too, the allusion here is to a *silver* ledge of considerable value; the mountains would later be known for their famous *gold* mine. Here are hints of death, and of Indians, and rich ore—all of which later played an important part in the Lost Dutchman legend. Indeed, all necessary ingredients are already here—and blended with enough passage of time, they make for interesting speculation.

If, perchance, one of the Germans was Waltz, this could account for a part of the legend that was to come, for Waltz was to have made a very rich find in the Superstitions, and $3,000 per ton ore was very rich, even by today's inflated standards. Given that one of the prospectors was Waltz, it is easy to visualize him in later years alluding to his Superstition "find," the mutilated body no doubt adding zest to his account.

On March 22, 1878, a little over a month after the previous article appeared in ***The Arizona Citizen***, a weekly Florence newspaper published the following account under its heading of "Mining Locations:" a claim named "Malakoff, in the Superstition Mountains," had recently been filed "by Oscar Macholz, John Wedgelich, and C. Smeitt."

Could these have been the Germans of the prior article? For sure, these names had strong Germanic overtones and the timing of the claim registration was about right. But, the article had said that two Germans had arrived in Florence and not three, leaving the tantalizing possibility open that one could have been the Dutchman. If Waltz was one of the two nameless men, he didn't bother to file a claim on the discovery, an action we will see later that was incongruous with the Dutchman's character.

But there was no doubt now that the Superstitions was a place to prospect for gold and silver, and when the paper said that "Many are preparing to visit the new silver fields," it couldn't have foretold better, for more people have prospected the Superstitions than any area on earth of similar size.

Two years later on Friday, December 5, 1879, **The Phoenix Herald** relates that a "Startling Strike near Superstition Mountain"had been made.

Some excitement is being created among the Mexican population of Phoenix by the story of a Mexican who arrived last evening from the Reno Mountains. He came into town under cover of darkness, as he was nearly naked. His hands and feet were torn and bloody, and his face was gashed in a fearful manner. His story is told with the air of a man who had been terribly frightened and had not recovered. With a companion he had started out prospecting about a month ago, going up the Salt River. They left the river when opposite the Superstition Mountain. Their prospecting began at this point. While climbing up the mountain, in a little gully, through black sand, and down which a large stream of water had evidently passed years ago, they were astonished to find that in this sand were large quantities of fine gold. In some places the sand was only about half an inch deep over the granite. The gold in pieces the size of a bean and smaller, was found in the little fissures in the face of the bedrock. Very little washing was necessary, and they found a little spring of water which furnished them what they needed. They obtained, they think, about $600 worth in half a day's work. About 2 o'clock in the afternoon, they were surprised to see an Indian woman come to the top of the gulch above the spring and start to come down. Upon seeing them she ran back over the hill. In less than ten minutes they were surrounded by fifty or sixty savages. The Indians were very small and seemed to be of a different nation than they had ever seen in Arizona. The Mexicans were not armed except with knives, and the purveyor says they were almost instantly caught with lariats. The Indians took them up the mountain and put them in a cave. They tormented and killed his companion and his fate would have been the same but for his escape. He succeeded in getting away with only a few knife gashes on his face. They lost their gold with all their outfit. The Indians seemed to be cave dwellers, and were evidently excited over the place being found by outsiders. Our reporter's limited knowledge of the Spanish language makes it impossible for us to obtain all the particulars of the affair. For the benefit of non-residents, we will say that Superstition Mountain derives its name from the fact that no *white man has ever been seen again, who attempted its ascension. It is a tradition among the Mexicans that large deposits of free gold are to be found in its gulches and ravines. It is not known whether there is any water there or not. We shall endeavor to obtain further particulars regarding the matter and will publish them as soon as obtained.*

Although nothing more was written about that particular event, it was not forgotten. Gold for the taking, Indians of a type never before seen, torture, death, and escape. Old Superstition already had a reputation and it was getting better as the years passed. A hint is given here as to how Superstition Mountain may have received its name: no white man who attempted to climb it was ever seen again. Pretty strong medicine admittedly, but Superstition's reputation would get worse. Prospectors, Indians, gold, and death; ingredients that if mixed right, would give birth to the world's greatest lost mine legend.

Continuing the journey by newspapers, the calendar is turned ahead a few years from the previous article to June 19, 1884, and another Phoenix paper, the *Arizona Gazette*. Here, in one of the rare instances it can be found in print, Jacob Waltz' name is mentioned; and strangely enough, it is associated with a murder. Old Jacob was home that night when the killing took place, but it was not until:

> *...About 10 o'clock a.m. yesterday word was received that a Mexican by the name of Pedro Ortega had been murdered, at the house of Jacob Waltz, a mile southeast of this city, by a man named Selso Grajalva. The matter was turned over to Justice Richards, who, acting as a coroner, summoned a jury and repaired to the scene of the tragedy at 1 o'clock. The Gazette reporter was on hand. He found the body of Ortega lying about thirty feet from the house of Jacob Waltz, his legs and abdomen from the knees to the breast bone were perforated with "double B" shot, the femoral artery on the right side being severed and in itself producing death. There is a mystery about the whole affair. Ortega was shot and killed by a shotgun belonging to Jacob Waltz. That gentleman heard the Mexicans talking loudly, and then the report of the gun. He ran to the side of the house where the tragedy occurred and saw the dying man. The assassin evidently ran around the building in an opposite direction, returned the gun to the room and place from whence it had been taken, and then sought flight. Waltz did not know, and so testified at the inquest, that his gun had been the instrument of the crime until Deputy Sheriff Rogers appeared on the scene. The only direct testimony was that of Waltz....The murderer is still at large, although officers are in hot pursuit.*

Who was the real murderer? Was it Selso Grajalva as the writer claimed, or had Jacob Waltz found the Mexicans on his property looking for gold he supposedly had hidden and shot one of them? As

Waltz was the only person on the scene at the time of the murder, his testimony could not be doubted; and too, the murder of a Mexican in those days was not looked upon as being quite as serious as killing a white man. No follow-up news on the murder could be found, so only speculation can persist about the real murderer. Perhaps Waltz did shoot Ortega, and Grajalva, sensing his life was at stake, fled. There were no witnesses, and only the testimony of Jacob Waltz stands as the record of the event. However, he *could* have lied. It would have been easy to do and there was no way to contradict his word. If the two Mexicans discovered something Waltz didn't want known, the easiest way to silence them would have been by using the loaded shotgun he kept nearby. But, the foregoing refers to Jacob as a "gentleman," and even though his gun was employed as the murder instrument, there was not a mention of him as a possible suspect. His word in this case seemed to be sufficient to place the blame on the other Mexican. The coming years would not be so kind to Waltz' reputation, for he would be accused of killing many persons in his quest for Superstition's gold. Many of the stories would picture him as a blood-thirsty killer who would not hesitate to shoot anyone who approached his mining claims in the mysterious mountains—and the number of persons who did so was considerable.

That same year the ***Phoenix Herald*** again mentions the Superstition Mountains, adding additional details of the mountains' mysterious faces:

> Messers Asa and Babcock of Tempe have been prospecting for some time in the Superstition Mountains, and report the discovery of 200 pre-historic war clubs. These clubs are of hard wood, ranging from 15-20 inches in length, with a grip carved on one end somewhat after the style of a policeman's club, and the other end is perhaps 3 inches in diameter.
> They were painted or carved with crude diagrams.

A few years later in 1886, the ***Phoenix Herald*** had a short article that told of Superstition's growing reputation. The notion that anyone who enters the mountains alone would not come out alive had already taken hold. Indeed, the Superstitions had been revered by nearby Indian tribes as a mysterious and foreboding place, and this awe and distrust was being passed on to the white settlers. The article's headline—itself shrouded in mystery—reads: "A Letter from ? in the Superstition Mountains." It goes on to say that:

> Strange stories have been circulating concerning mysterious disappearances of luckless prospectors who entered the range on an organized search for fabulous deposits of rich ore, and it is even asserted that quite a party of prospectors who entered the range on an organized search for the mother ledge of the famous Silver King were never heard of after. The Indians hold these mountains in

superstitious awe, and in fact all these myths and fables originated with the Pima Indians many years ago.

The Superstitions claimed its victims then as it does today. Prospectors still search for her hidden gold and they still die. The mountains never seem to tire of their duty.

One of the last mentions of Jacob Waltz is a somewhat trivial notice that appeared in the February 26, 1891, issue of the ***Weekly Phoenix Herald***. It concerned the disastrous flood that hit Phoenix only a short time before:

> *Ed Scarborough, Henry King and others started about 9 p.m. with a boat to rescue Count Duppa, H. Tweed and Jake Star, at the latter's place southeast of town. They were on the roof but unconcerned, as a levee thrown up around the adobe house rendered it proof against the flood. All three men appeared in town this morning as though nothing had happened. Near their places several adobes, not having been embanked, fell, including those of Beckett, Pesquica, butcher Grijalba, Jake Walts, one belonging to four Sweeds, and two Mexican houses back of Starr's.*

Flood indeed! That February of 1891, the city of Phoenix was victim of one of nature's strangest and most unpredictable maladies, and also one of her most destructive—the flooding of a desert river. The Salt River, fed by storms in the mountains, took its revenge, unleashing its waters, causing the most extensive damage ever suffered by Phoenix and neighboring Tempe.

Jacob Waltz had been a victim of that flood. His house was toppled and he had been caught in the torrential waters. Many of his possessions were swept away, leaving him dependent upon his friends. Shortly afterward Waltz came down with pneumonia, leaving him entirely under the watch of Julia Thomas, a black woman whom he had earlier befriended. The "Jake Walts" of the above article was Jacob Waltz, owner of a small adobe house which was not embanked and was laid to waste by the ravaging flood. This is the same man who the ensuing years would label as having owned one of the world's richest gold mines. But he didn't seem to have enough money in 1891 to have his humble house embanked against the flood!

Old Jacob had but a short time to live and the few remaining months were spent in bed. He relied on a handful of friends to nurse him through the final days; and on October 25, 1891, he succumbed to complications of pneumonia that had gripped him some eight months prior. In writing his epitaph the ***Phoenix Herald*** said: "Jacob Waltz, aged 81 years, died at 6:00 a.m. Sunday, October 25, 1891, and was buried at ten o'clock this morning from the residence of Mrs. J. E. Thomas, who had kindly nursed him through his last illness. Deceased was a native of Germany and spent the last thirty years of his life in

Arizona, mining part of the time, ranching and raising chickens. His honest, amiable, industrious character led Mrs. Thomas to care for him during his final days on earth and he died with a blessing for her on his lips."

The *Arizona Daily Gazette* was not so kind when it presented its terse notice that: "Jacob Walts died Sunday evening at the residence of Mrs. J. E. Thomas and was buried yesterday. Deceased was a native of Germany and was 81 years old."

That's all—no more, no less; the final public tribute to the man who would become the central figure of one of the world's best-known legends. It is worthy to note that the *Herald* describes Waltz as being of "honest, amiable, industrious character"; it also makes note of the fact that he did devote some time to mining; however, there is no mention of him owning any type of mine in the Superstitions. With the death of Jacob Waltz, the legendary Waltz was born, the Dutchman—owner of the richest mine in the world—and this time it would take more than pneumonia to fell him.

ROCK WRITINGS
Indian tales written on rockside at Hieroglyphic Canyon.

"...twirling...like a spider... on the end of a thread."

vignette 5

The sun had yet to burn its way through the Superstition early morning mist when Louie yelled in the direction of my tent, "Breakfast is almost ready!"

I was anxious to get going, for last night Louie told me we would be climbing to the top of Weaver's Needle. It looked like a formidable task, but I felt I was ready for it. My curiosity and excitement had been fired that previous night as I learned that somewhere on the Needle was where Maria thought the lost gold was located.

After a large breakfast, we packed our gear—water, food and guns—along with a few hundred feet of nylon rope Louie had purchased on his last trip to Phoenix. Maria, dressed in her usual mountain outfit of khaki pants, old shirt, wide-brimmed hat, sunglasses and the ever-present tennis shoes, would carry only water and food for herself. A pistol hung at her side.

A short climb took us to the base of the main upthrust of Weaver's Needle, the point where the main dike or plug was pushed upward some 1,000 feet. Approaching the main spire, I could see where someone had already fastened ropes to the inside of a chimney or vertical opening that split the Needle on the south side, making a kind of makeshift elevator shaft up through the center of the Needle. Louie said he had attached the ropes on previous climbs and that they made the job of getting to the top much easier. He instructed me to follow his lead and grab where he grabbed and step where he stepped. He knew the Needle well and had climbed it many times. Maria would wait at the point where the ropes began. She would go no further. I couldn't visualize her climbing, anyway, and looking at the chore that lay before us, I couldn't blame her for staying put.

"Let's go!" Louie said, pointing his finger to the sky.

Using the ropes that had already been placed, we started up. Louie, who couldn't have weighed more than 125 pounds, said he would carry the extra rope. I didn't argue.

The first part of the climb was much like going up the inside of a narrow well where one can push both ways with his arms and then hold that position while the feet are lifted to the next secure footing. This "well" however, had a rope hanging down it which helped in pulling up to the next toe hold. As I climbed I wondered how well the ropes were attached above and placed more emphasis on my own hand and foot holds than I did on the rope. I mused that we were literally "threading the needle."

"Louie, wait up" I yelled. No answer. He had already scrambled out of sight somewhere above me, forgetting apparently his advice that I should grab where he grabbed and step where he stepped. I couldn't grab or step where he did if I couldn't see him. I shouted again.

"Loooouie!"

No reply. I would just have to do it on my own. Carefully, foot by foot, not really trusting the rope, I worked my way up. About half way, I ran into Louie checking the securing point of the rope we had been using. It was wrapped around a large boulder, but I noticed that at the point it dropped off the ledge we were on it looked a bit frayed. I was glad I didn't place too much faith in it.

"The hard part is over," Louie said, "we can make the rest of the way without ropes."

As we climbed, the wind began to pick up. I wondered what it would be like at the top. I pondered, too, the extra length of rope Louie had brought. He said it would be used when we reached the top—I could only guess for what.

The remainder of the climb went smoothly and I was surprised to see that the top of Weaver's Needle consisted of a relatively broad expanse of fairly flat terrain. The view from the top was quite fantastic, with seemingly unlimited visibility to the north and the east. While I was walking around getting acquainted with the topography, an Air Force jet buzzed by, wobbling its wings in salute as it passed a few hundred feet overhead.

I followed Louie to the east side where it is possible to descend quite some distance down a gentle slope. However, at one point the terrain abruptly drops off and then breaks in toward the west, so that looking straight down from that point you can't see much of the east face of the Needle, only the ground some 1,000 feet below. I swallowed as I saw Louie hooking the rope he had brought to a boulder not far from the overhang. So that's what the rope is for, I thought. We're going to lower ourselves over the ledge.

Louie said it would be easy and it would be the best way to get a bird's-eye view of part of the east face of the Needle, the place Maria had said there would be some type of opening into the Needle itself. Maria was convinced that the treasure of the Superstitions was hidden somewhere here, high above the desert floor. She believed there was an opening, be it a door, a cave or a crevice in the east face of the spire, and it was supposed to be well hidden. Behind that hidden entrance in a large hollowed-out room or cave, deep inside the Needle, lay the lost Superstition gold. The upshot of her being in the mountains was to find that opening.

Louie's idea was to lower the rope and let it dangle for a few hundred feet, then we could shimmy down the rope to a point near the end where he had tied a large knot. By holding the rope and standing on the knot we could start the rope swinging back and forth like a pendulum, until the arc was sufficient to swing us over to the side of the Needle where, landing site permitting, we could let go and drop to solid ground. A small cord would link the jumper with the main rope, so that when it was time to go back up, you could retrieve the main line by pulling on the cord.

"You're crazy," I said to him.

"It should work," Louie replied. I didn't like him using "should."

I talked him into tying knots in the rope at two foot intervals. This

LOUIE AND AUTHOR
Photo of Louie (left) and Robert Sikorsky near their camp.

Photo taken by Maria Jones

should make it much easier to come up and down, I reasoned.

"Let's go," said Louie, stepping back and grabbing the rope near the boulder. "I thought that since you were the heaviest it would be better for you to go first. I can hold the rope and make sure it is OK."

My reasoning was just the opposite—the lightest one should go first. However, it didn't work out that way. I went first.

Ever so slowly, ever so carefully, I took my first tentative steps into space, locking my legs, shoes and hands on each knot as I descended...I dared not look down and kept my eyes on the rope in front of me.

Then something happened that neither of us had counted on. The rope started to spin, twirling me around and around like a spider on the end of a long thread. First this way, then that way, aided no doubt by the strong breeze that was blowing. My stomach tightened. I felt completely helpless and disoriented. I could hear Louie yelling something from above, but my only concern now was to stop myself from spinning. I began to pump back and forth trying to get the rope and myself moving in an east-west direction toward the Needle. This succeeded somewhat in stopping the spinning, but each time I tried to lower myself a bit, the spinning would start again. We should have a ladder rope, I thought. After many tries, I finally learned how to adjust my body weight to counteract the spinning effect and slowly edged down to the end of the rope. The vertical cliff face lay a mere 30 feet to the east and, as I scoured its features, I could see no place a person could find footing. There was, however, a small ledge about 100 feet below, but that would mean going back up and leaving out more rope. Yelling to Louie was useless. The wind and the tons of rock between us cut off any effort at communication.

I looked closely at the face of the cliff. There was nothing that even resembled a door or cave or opening into the mountain, but then, my viewing was limited to some 150 feet of vertical distance. It could be below, near the ledge I had spotted.

There was nothing more I could do, I had done my job. Slowly, easily, I began inching my way back up the rope, grateful to myself for having suggested tying those knots every few feet, for they were like little islands of security each time I reached one. I became acutely aware of the half-inch nylon rope that stood between me and a thousand foot fall. I wanted off that rope. Looking up I could see Louie peering over the ledge, grinning, chattering excitedly. I made it back, silently thankful as I grabbed his hand and he helped pull me the last few feet. Solid rock never felt so good.

"Did you swing over to the face of the cliff?" Louie anxiously inquired.

"No, there was no place for me to land, so I couldn't see swinging back and forth for no reason. There is a ledge about 100 feet lower and I think it could easily hold a person."

"That must be the spot Maria means," Louie offered. "She said there is a ledge not far down from the top. Let's lower more rope and see if we can make it this time."

"No way I'm going down again," I said. "I still have the shakes. It's your baby Louie, I'll stay and watch the rope this time."

Curiously, he decided that we had had enough for one day and suggested we start back down, as the hour was getting late and Maria would begin to worry if we weren't back soon. I didn't argue. Louie led the way and within minutes was out of sight. I followed, not really sure of what I was doing there, not yet aware of the danger I had just placed myself in.

As I continued my descent I heard gun shots coming from below. One, two, three...a total of six shots in all. Was Maria or Louie in trouble? I quickened my pace, prodded by the fear that was knotting my stomach. What if someone had killed them? Flickering thoughts of the many people who had died in the Superstitions raced through my mind. Careful, careful, I thought to myself as I raced down between the narrow walls of the chimney. Easy does it, if someone is down there they probably don't know I'm still up here.

About a hundred feet from bottom, I could hear voices. I came to a dead stop, every muscle in my body tense, straining to hold myself still, senses alert, trying to pick up fragments of the conversation going on below. It was hard to hear over the wind that whistled up through the Needle. I dared not move, not wanting to disturb the ground and send a rock tumbling down, a sure signal of my presence. Then my ears picked up a familiar sound, the loud, happy, carefree chortling of Maria. She was laughing about something. Relieved, I scurried down the remainder of the chimney. Louie and Maria were sitting there, chatting happily, munching on apples and drinking water.

"What happened? What was all the shooting about?" I asked.

Maria looked at me sheepishly. "I saw a Gila monster right there in those rocks and emptied my pistol at him," she drawled.

I looked at Louie. He took another bite of his apple and shrugged his shoulders. He was used to it.

5
Jacob Waltz

What little is known today of Jacob Waltz has been handed down through three sources: the newspapers, legal documents, and word of mouth. Of the three, the last named is responsible for most of our knowledge but at the same time is to blame for the maze of contradictions and uncertainties that surround Waltz. Newspaper accounts and legal documents, on the other hand, supply some touch, vague as it may be, with reality and truth, and act as watchmen in setting certain limits to the amount and direction of speculation about old Jacob. In this manner, Waltz's character and the legend of his lost mine will always remain within fixed borders, constantly held in check by the scanty historical record.

Jacob Waltz was born in Prussia, most likely in the year 1810 (the exact date of his birth is unknown). Speculation had it that in the year 1839 or thereabouts, Waltz landed in New York harbor, and set foot upon American soil for the first time. From New York he journeyed to St. Louis, most probably settling near people of similar backgrounds. What happened next is not known, but eventually Waltz headed west, lured by the promise of the California gold fields.

The picture Jacob Waltz presents during his early years in the United States is, at best, a cloudy one. It is not until he came to Arizona that the veil lifts a bit. He allegedly worked around San Francisco a number of years and also tried his luck in the California mother lode country. Mexico, especially its gold-bearing provinces to the north, possibly claimed some of the Dutchman's time. We can trace his California movements somewhat by turning to a census dated October 27, 1850, taken for the City and County of Sacramento, California. Here listed as a resident of the County is a J. W. Walls, possibly the Jacob Waltz of Dutchman fame. If this was **the** Jacob Waltz, he was probably attracted to the area, along with the many other 49'ers, hoping to strike it rich in the mother lode. The Dutchman, now turned forty, was searching for that nest egg that so far had eluded him. Ten years later we find Waltz listed in the Los Angeles County Census of July 24, 1860. He is a resident of the Los Angeles area, employed in a trade that would later be his Arizona trademark—mining. It is unfortunate that we do not know more about Waltz's early life, but if we did, he would no longer be the mysterious character portrayed in the legend. You may have read this and that about Waltz because authors have written what they claim is the "true" story of his life. They have presented "proof" for their allegations, but the "proof" has fallen far short. There is no way to prove word-of-mouth information that has been passed through the years. The uncertainties and discrepancies that have crept into the spotlight are mostly the work of men's flexible imaginations, and are to blame for the legendary Jacob Waltz of today.

But of the following we **are** sure: on July 19, 1861, in the Court for the first District, Los Angeles, California, Jacob Waltz was naturalized as a citizen of the United States. Approximately twenty-two years had transpired from the time Waltz entered the United States to the time he became a citizen. During these "blank years" Waltz was unravelling the many wonders of his new-found land, feeling his way through a variety of jobs and enjoying the freedom. That Waltz was interested enough to become a citizen, hints perhaps, at his early character.

Jacob Waltz arrived in Arizona in 1862, just a few short months after he was naturalized. He was 52 years old. As proof of his residence in Arizona we can turn to the Territorial and U.S. Census as well as the Great Register of Maricopa County. In the Territorial Census of April, 1864, Third District, Yavapai County, he is listed as Jacob Waltz (note the spelling), age 54, birthplace—Germany, occupation miner, resident of Arizona for two years. His name appears on the Great Register of Maricopa County (in which Phoenix is located) for the years 1876, 1882, and 1886, listed as a resident of Phoenix, born in Germany, with ages of 66, 72, and 76, respectively. He is again recorded as Jacob Waltz in the United States Census taken at Phoenix in 1880, age 70, born in Prussia, and occupation, of all things, **farmer!**

Over the years there has been much disagreement about the correct spelling of Jacob's name. He has been called Waltz, Woltz, Wolz, Walts, Wals, Waltzer, von Walzer, and Walzer; and in a way, all of these are correct! The name Jacob used himself and the one he **signed** was **Waltz.** It is very possible that his name at birth **was** Waltzer or von Waltzer, as is claimed; but upon arrival in America he shortened it to Waltz. Phonetically, "Waltz" was easier, and in those days people had a tendency to call a man by the simplest way. In almost all historical accounts the spelling of his name is given as Waltz. Any discrepancies in the spelling can be accounted for in the close phonetic similarity between the name Waltz and its counterparts.

Another dispute revolves around Jacob Waltz's intelligence. Most have conceded that he was shrewd and cunning—the traits so often associated with him; others have said that he couldn't read or write and that he spoke English haltingly; and on the extreme are claims that Waltz graduated with a degree in mining engineering from Heidelberg University, one of Germany's oldest and most highly regarded institutes of learning. The last claim intrigued me and I inquired whether or not Jacob Waltz had at some time attended the university. In reply, a letter was received from the Registrar's office at Heidelberg saying that "we regret to tell you in answer to your letter... that we tried to trace a Mr. Jacob von Walzer or Jacob Waltz in the archives of this University. According to our files he was never enrolled at this University."

Although not a university graduate, Waltz nevertheless was not devoid of intellect and what written accounts we have show him as a thoughtful, conscientious person of above average intelligence. The legend, however, would not picture him that way.

In September, 1863, Jacob Waltz staked and recorded his first mining claim in Arizona. It was not anywhere near the Superstition Mountains but lay in the general area of Prescott, the then capital of the Territory. A short time later, in March 1864, Waltz, still probably working in the same region, affixed his signature to a petition, along with other "Citizens of Randalls District on the Haciamp Creek," asking Governor John N. Goodwin for protection against the Indians. The petitioners, mostly miners and farmers, wrote that:

> ...we occupy a District rich in Mineral Wealth and agricultural facilities in which much labor has been done and much money and energy expended in opening claims and making preparations for mining and farming. Our district is isolated from other settlements and nearer to the Indian Territory than any in the Mining Region. And the centre of our settlement is at the point at which the Trail from the Indian Country converges. A strong effort is being made by enterprising citizens to make this one of the permanent settlements of the Territory. Yet we are so constantly harassed by the Indians while at home and in such imminent danger of being attacked by them when we leave for supplies that it is almost impossible to carry on the work already commenced. On the 2nd of March the Indians visited the Settlement and killed eight men, so alarming our citizens as to cause some to abandon their claims entirely and leave the Territory, and the others to collect at one point for mutual protection, thereby leaving their work and their means of subsistence.
>
> We applied to your Excellency and were very promptly given a temporary relief. The time for which the Troops who came were (sic) rationed has now expired and many miners are preparing to leave the District, which will so weaken us who remain as to imperil the life of all who remain in the District. We would therefore ask that you station a small force in our midst, from twenty to thirty troops, Infantry, to protect the miners and establish a point of rendevous in case of an attack until we can get fairly under way and be receiving some income from the work so mostly completed. When we will be prepared and able not only to defend ourselves but to make formidable campaigns against the Indians.

Waltz had been working in an area that was "...isolated from other settlements and nearer to the Indian Territory than any in the Mining Region." Perhaps it was his later recollections of these contacts with the Indians and close calls with death that somehow wormed their way, via embellished accounts of his contemporaries, into the Dutchman legend. Indians and death made fascinating conversation, even in those days.

Six months later, on September 14, 1864, Waltz staked the "Big Rebel" claim in the Walnut Grove Mining District in Yavapai County. He duly registered it on January 7th of the following year. On January 8th, just a day after registering the "Big Rebel," he filed on another claim, the "General Grant," in the same Recorder's office in Prescott. These two claims, along with the earlier one of September, 1863, constitute the known entirety of the Dutchman's "recorded" and verified mining activity. There is no known record of him ever having a "legal" claim in the Superstitions.

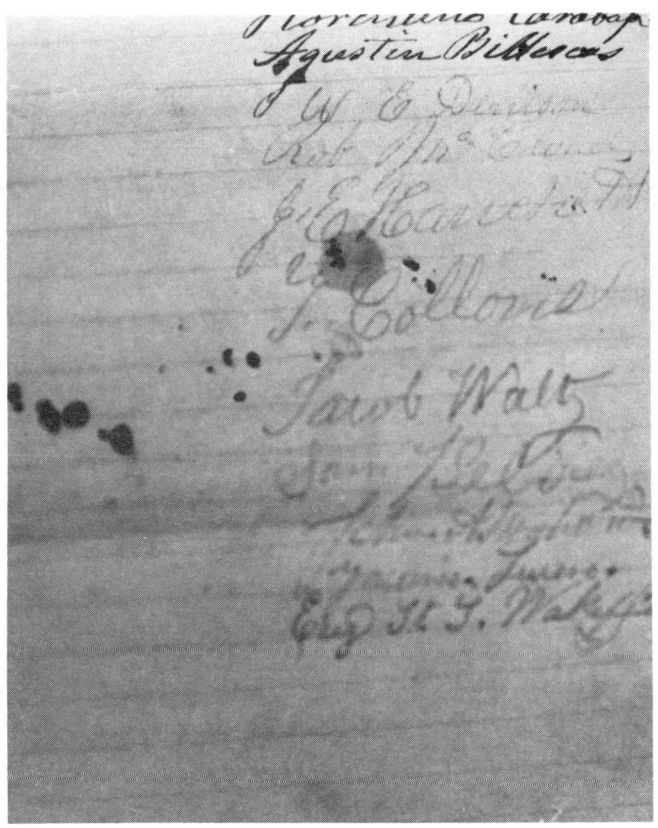

WALTZ'S HANDWRITING
The signature of Jacob Waltz appears, in pencil, on a petition to Gov. John N. Goodwin, asking for protection from marauding Indians. One of the other signers of this document was Henry Wickenburg, founder of Wickenburg, Arizona, and discoverer of the famous Vulture Mine.

By 1878 Waltz had been in Arizona for sixteen years—prospecting, mining, ranching and farming—eking out a living along with other pioneers, in the best way he knew. It would be interesting if we had some type of substantial evidence of his financial well-being at this time, for the legend says he discovered his mine some time in the 1870's. It was during these years that he is supposed to have set Phoenix and surrounding communities afire with lavish spending and drinking—and bragging of his gold mine. Fortunately, such a document **does** exist, and it gives—quite clearly—the financial status of Waltz.

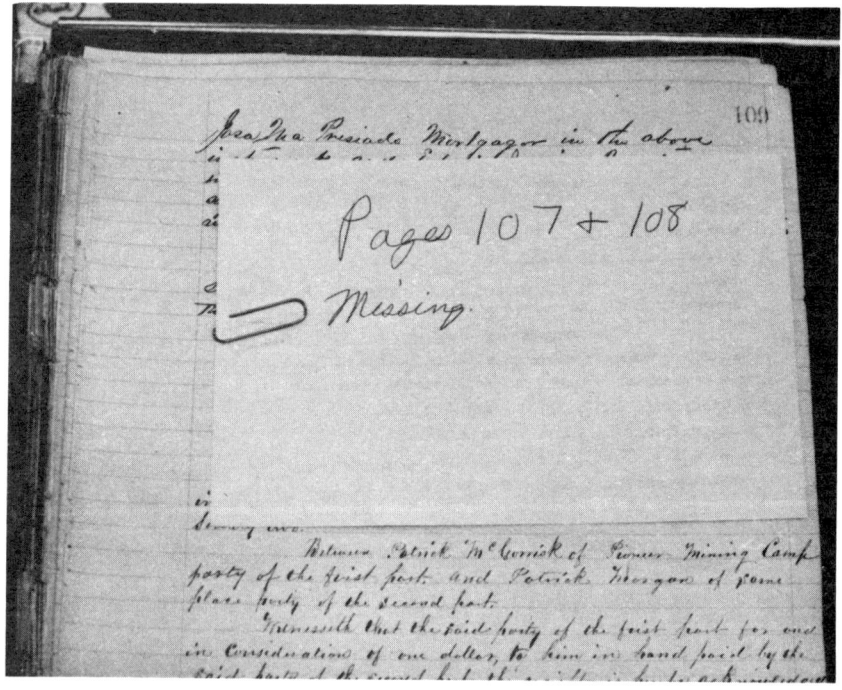

MISSING AFFIDAVIT
Page 107, Book 1, of Miscellaneous Records at the Maricopa County Recorder's Office is supposed to contain an affidavit of Jacob Waltz. Instead, this terse note greets the researcher. Someone, for reasons unknown, has ripped out and stolen the pages where Waltz's affidavit once was recorded.

"...not me, Maria."

vignette 6

One day, shortly after Maria and Raymond had returned from Phoenix where they had gone to pick up another grubstake from her mysterious benefactor (I never did find out where she got the funds to keep her operation going), I had a long chat with Maria. Now that she was flush again, I tried to talk her into hiring a helicopter to take close-up pictures of the side of Weaver's Needle she thought might have an opening in it. This was the same place that just a few short days ago I had inspected while on the end of a half-inch nylon rope, the place where she insisted we would find the door to the interior of the gigantic peak, the place that contained Superstition's gold. It seemed an odd place, geologically speaking, for any type of natural gold deposit to occur, in the center of a volcanic dike or plug. But Maria had hedged her bet. If a natural deposit wasn't there, then it would be a large storage room, a place big enough to hold a large portable bonanza. In fact, she even alluded to this being the place Montezuma may have hid his treasure after fleeing from the Spanish. It was virtually inaccessible, an excellent place to hide something if you wanted to go through the trouble of hiding it. But, try as I might, I couldn't visualize the Dutchman rappelling up and down the Needle after a long arduous trek from Phoenix.

I argued that we could hire the helicopter to go in and we could shoot pictures in the early morning, when the sun illuminated the eastern face. We could then have the pictures enlarged and scrutinize them for any sign of an opening or intrusion. This, I reasoned, would be a good investment and would save much time and be a lot less dangerous than hanging from a rope 1,000 feet from nowhere. But, she wouldn't buy my idea. For some reason—perhaps she was afraid that the photos would shatter the strands of the story she clung to—she was adamant in her refusal to go along with my way of thinking, even though there was a helicopter service available in Mesa, just some 15 miles east of the Superstitions. We would climb the Needle and lower ourselves to the ledge I had spotted. There, she insisted, the opening would be. There was no need to hire a helicopter.

You, or Louie, or Raymond might lower yourselves, I thought, as our conversation ended, but not me, Maria.

6
The Dutchman's Darkest Day

Dawn of August 8, 1878, brought the great furnace of heat so common to Phoenix in the late summer months. By noon the temperature had risen over the hundred mark and humidity still present from the night's rain readily coaxed sweat from anyone who dared chance activity. Not bothered by the searing heat, a man sat alone in his small adobe house which lay humbly on the village outskirts. Laboriously and with unpracticed hand he wrote. He was not used to writing nor to the fancy legal jargon he was trying to phrase. A few days earlier, a friend had given him a rough draft of the document and offered to finish it, but he doggedly insisted on doing it himself.

Outside, chickens scrambled for bits of corn left over from the morning's feeding, and a solitary cow stood placidly lolling in the skimpy shade offered by a palo verde tree. There wasn't much here on this quarter section of land—a small, precariously built adobe house, a few chickens, a cow, and scattered odds and ends of mining and prospecting equipment. A horse and wagon stood outside the gate of the crude picket fence that surrounded the house.

Inside, the obvious ineptness of the man's writing was punctuated by occasional bursts of cursing followed by the paper being ripped up and another started. Finally, at 4 p.m., in the blazing cauldron of the afternoon's heat, he finished. He had to meet Andy Starar at 4:30 and if he didn't hurry he would be late. He carelessly splashed water on his face, more to cool him than to clean, put on some clean clothes, brushed his fine long beard, and hopped into the waiting wagon.

Arriving at the courthouse at exactly 4:30, he saw Andy Starar and they met with a brief handshake. To Starar, the man seemed very despondent and extremely moody, and his attempts to cheer him up were met with curt rebuffs. They were friends, however, and Starar understood the situation.

Indeed, Jacob Waltz was at a very low ebb. His 68 years weighed heavily and there was a sad, foreboding look in his eyes. He had come to a dead end. Beaten and anxiety-ridden, he called upon one of his few friends for help, and Andy Starar suggested that old Jacob draw up the document. Jacob readily agreed, exacting the contents of the paper himself. They met at the courthouse to sign and make the paper official. Waltz's sullen eagerness was evident—it would lift a great burden from his already stooped and beaten shoulders. Yes, Jacob Waltz was the man who, unknowingly at the time, was destined for immortality. It was this man who, according to legend, discovered the fabulous gold mine in the Superstition Mountains and set Phoenix into a frenzy with his ribald spending and drinking. Yet it was he who drew up and signed the document which, recorded at 5 p.m., August 8, 1878, on pages 322-325 of Book Three of Deeds, Maricopa County Courthouse, Phoenix, Arizona, read as follows:

> This agreement made and entered into this 8th day of August in the year of our Lord 1878—between Jacob Waltz...and Andrew Starar...Witnesseth that, Jacob Waltz, for and in consideration of the promises and covenants of Andrew Starar,...and the further consideration of the sum of fifty dollars to him in hand paid by Andrew Starar...hath sold, granted, confirmed, remised and released to Andrew Starar the following property, real and personal, to wit: all the right title and interest of Jacob Waltz in and to the north east quarter of Section 16 in Township 1 north, of Range 3 east... With all the improvements and appurtenances belonging or in any wise appertaining. Also all the grain now due and owing from the estate of Ferdinand Madgeburg deceased...Also all the chickens on said ranch or land..., also two horses now owned by Jacob Waltz and all other personal property of every kind and description now owned by Jacob Waltz, whether legal, equitable, whether real, personal or mixed, giving to Andrew Starar full possession and property in all of the said estate and property above described.
>
> And for and in consideration of the said above described property and possession of the same, Andrew Starar agrees and promises that during the full continuance of the natural life of Jacob Waltz and to the end thereof, he, Andrew Starar, will furnish to Jacob Waltz all necessary food and clothing required, together with protection and all necessary medecines (sic) and attendance when sick. He, Andrew Starar... shall take such care of Jacob Waltz as shall be necessary and proper, in degree as one of the family... The interest and full meaning of this agreement is that Jacob Waltz shall make over and convey all of his property of every kind to Andrew Starar, and that Andrew Starar shall be full owner of the same, without any recourse or return to Jacob Waltz, and by these presents Jacob Waltz does make over, transfer and convey to Andrew Starar, all of said property, and in consideration thereof, Andrew Starar shall take care of, feed, clothe and protect Jacob Waltz in a good and proper manner suitable to his ordinary style of life, and this Andrew Starar by these presents agrees to do.
>
> In witness whereof we have hereunto set our hands and seals, both Jacob Waltz and Andrew Starar..."
>
> Signed: **Jacob Waltz**
Andrew Starar

(Note: For ease of reading, the names of Jacob Waltz and Andrew Starar have been substituted in the above document in place of the actual legal terminology used—"the said party of the first part and the said party of the second part.")

In affixing his official seal to the document, John T. Alsap, the Notary Public who transcribed it, noted that "each of them acknowledged to me that they executed (the agreement) freely and voluntarily..."

DUTCHMAN'S DECISION
This is the final page of the agreement between Jacob Waltz and Andrew Starar referred to in the "Dutchman's Darkest Day" chapter.

Jacob Waltz—possessor of the secret to the richest mine in the world—had just signed a document that, in effect, made him the ward of another man! His property was very little—a quarter-section of land and a small adobe house were the main attributes. He was probably broke, for when Andrew Starar signed the document and became full owner of all of Dutch Jacob's belongings, he forwarded Jacob fifty dollars. If Waltz knew the location of the fabulous mine, he showed no signs of it. He seemed to have lost all faith in himself and his abilities and he willingly signed away his property so Andy Starar would feed, clothe, and protect him, take care of him when he was sick, and treat him in the same degree as one of his own family.

Andrew and Jacob Starar had a large ranch on the outskirts of

Phoenix and the German-born brothers were prominent in city affairs. Waltz became acquainted with them through their common bond of language and like interests in civic affairs. It was not unusual in those days, as now, for someone in need to turn to their own kind; thus Waltz, knowing that a person of German background could be trusted, sought the Starars in his darkest hour.

The seventeen years Waltz had spent in Arizona seemed fruitless. He had little in way of material possessions, and years of prospecting and fighting the relentless desert had taken toll of his once robust countenance. If he had a mine or even a claim, he never bothered to register it legally, an unwise act in those days when prospecting was at its height of activity. If he knew the whereabouts of a cache of hidden gold, he balked at retrieving it. Waltz, at this stage of his life, was perfectly content to find security in the document he had signed.

It is at this point that we start to tread on the borderline between recorded history, folklore, and man's imagination. Dates become confused and are as real and meaningful as the individual permits. Fact and fiction become intertwined and man believes what he wants. Did Waltz have some kind of a mine or didn't he? At this point, the answer assuredly seems to be "no," but Waltz still had thirteen years left on this earth, and for a man who would become a legend, anything was still possible.

That Jacob Waltz had information concerning some type of mine or cache of gold in the Superstitions at the time of his death, and that he volunteered it to those at his death-bed, has been a lingering point of contention. Some say he never told a soul the exact location of his mine, mainly because he never had one; others claim he gave specific directions to the point of drawing a map. Was there, then, any basis for the story of a hidden mine in the Superstition Mountains? Indeed there **was!** Soon it will be shown that Waltz **did** divulge the location of his hidden bonanza in the Superstitions but that his description has proved to be quite inadequate.

vignette 7
"...slipped and fell some thousand odd feet..."

(A few years after I terminated my brief association with Maria Jones, I read in the papers that she had hired a Phoenix mining engineer, Vance Bacon, to investigate Weaver's Needle, presumably for signs of mineralization. While in the process of climbing the Needle, Bacon slipped and fell some thousand odd feet to his death. After reading about this unfortunate incident, I couldn't help but wonder if he had gone up the same way I had a number of years earlier, if the same old ropes were still there, if Louie had climbed with him and if he had fallen off the east face where I had dangled in space for a seeming eternity. Did he use the knotted rope? Did he fall as he was climbing up the center of the Needle or did he fall off the precipitous east side? Questions raced through my mind as I read the newspaper account. Was this the mining engineer I met and talked to briefly on my very last day of employment with Maria? I wondered at these things but really didn't want to know the answers.)

7
The "Superstitious" Indians

Perhaps there was no group, excluding the white man, that figured so mightily in the history of the western United States as the Indian. History is filled with stories of the Apache and other tribes. The mid and late 1800's were times in which white settlers and pioneers felt the full impact of the Indian, especially the Apache, and Arizona Territory was cruelly singled out. The contact between Indian and whites in this area alone is, in itself, a story that fills volumes.

Let's narrow the area of Apache habitation to a still smaller size—south central Arizona, and more specifically, the Superstition Mountains. The word "Superstition" carries many connotations, including its close association with the Apache and other Indian tribes. Mention the Superstition Mountains and you are sure to hear the word "Apache".

Legends, superstitions, myths, stories, old prospectors' tales—all tell of the Apache, hidden gold, and the Superstitions. Most of them are the product of the white man's over-taxed imagination, and very few can be traced to Apache origins.

It is a fact that Apaches **did** live in the Superstitions, and even today, remnants of them can be found in surrounding areas. Let's take a closer look to see who these people were and how they came to figure so heavily in the lore and legend of that fabled region.

The Apache Indians of the Superstition Mountain area actually may not be Apaches after all! Albert Schroeder, an anthropologist, says that they may have been Yavapai Indians, long time antagonists of the Apache. But either way, Apache or Yavapai, they meant the same—trouble for those who entered the Superstition range.

The word "Apache" however, had been misused over the years; the early Spaniards used it to sometimes mean "enemy" and other times referring to any hostile Indian group. To the Americans that followed, "Apache" meant many things, most importantly, bloodshed and misery. The Yavapai, who resembled the Apache in physical characteristics, could have easily been mistaken for their antagonists. But, all things considered, the name "Apache" came to be closely associated with the Superstitions.

Originally, the Yavapai and Apache were distinguishable by linguistic differences, the Apache being of Athabascan stock; over the years they have become intermingled, and although respective languages have not been seriously altered, the generations of mixed marriages have produced great confusion, rendering almost meaningless the concept of "pure blood".

Some Apaches claim they originally came from somewhere in the north and tribal stories hint vaguely of their "sister people." In fact, this may be very close to the truth, taking into account that the Navajo, who live to the north, are of the same stock. Other Apaches claim they are descendants of "people who live in the cliffs." The Superstitions are dotted with cliff dwellings and caves, and many show signs of former inhabitants. It has never been decided who these ancient dwellers were. Some guesses are that the people were what we call Hohokam or Salado, ancient ones who populated the Valley of the Sun and regions to the east of the valley many centuries ago.

The Superstition Mountains became a sanctuary for the constantly harassed Apache. General George Crook, the famed Indian fighter, sent many patrols after the Apaches. Time and again they would lose the elusive foe in the rugged mountains, for this had been the Apache's sacred land and they wanted no white man there.

In the year 1872 a great disaster befell the Apaches of the Superstition Mountain, throwing the balance of power to the white man. It is a tale of horror and death and is responsible for many of the other stories of death that have come out of the Superstitions. It was in that year the first organized effort was made to drive the Apaches from the mountains.

In December of 1872, a column of troops of the Fifth United States cavalry, with about one hundred Maricopa Indian scouts, came upon the trail of a large band of Indians in the western foothills of the Superstition Mountains. The trail was followed into the wildest part of the range and then was lost on a granite ridge, amid a maze of canyons—lost even to the sharpest among the Maricopas, reputed to be the most expert trackers in all America. Soon, however, the scouts discovered an Apache brave and a young Indian boy hiding in a rocky crevice. The elder escaped them, but the child was captured and forced to lead the way to the nearby tribal camp. Only a few hours before, he had left his parents to go with his uncle and scout for the approaching enemy.

The Apache hiding place was a natural fortress—an immense cave, with but a single entrance. It was located at the very lip of an impassable gorge which extended to become part of the great Salt River canyon. The only approach was by a very narrow trail, hemmed in on both sides by steep, rocky walls, making it practically impossible for large forces to approach the cave *en masse.* Inside the cave the band of approximately one hundred and fifty Apaches had found fairly comfortable, though crowded, quarters. At daybreak next morning the pursuing troops demanded surrender, only to be answered with jeers and defiance. The Apaches felt that their cave with its restricted entrance would secure them from the cavalry rifles. Direct firing into the cave was impossible because of the right angle turn the steep trail made as it approached the entrance. The troopers and scouts, from every possible vantage point along the gorge, opened fire on the cave. It only proved futile.

Soon, however, one of the Maricopa scouts discovered that a sloping granite slab overhung the entrance to the Apaches hideout, and bullets fired at the slab would deflect into the cave. In a moment his suggestion was taken advantage of and a hundred troopers began using the slab as a target. The huddled and confident Apaches were inundated with a rain of fifty-caliber bullets. In half an hour the commanding officer ordered a cease-fire and again demanded surrender. No answer was forthcoming from the cave and no sign of life was in evidence. A bold scout crept forward along the ledge. He heard a faint moaning and then, peeping around the angle, saw a terrible sight.

Of all the braves, squaws and children within, only a few seemed to have escaped the fusillade. The dead were in heaps and rows. Mother and child had died from the same bullet. The walls were scarred from the bullets' glances. The whoop of the scout brought the other Maricopas to him with a rush. With diabolical glee, they started to kill the few survivors and by the time the troopers appeared all had been killed.

This battle, or rather slaughter, became known as the "Battle of Skull Cave," and it is amazing that over thirty years passed before that gruesome Apache tomb became publicly known. Indians and death: two things that were to haunt the legend of the Lost Dutchman in the coming years.

The Indians of the Superstition Mountains had legends concerning that area long before the first *extranjeros* set foot there. Many of the more peaceful tribes that lived and farmed near the mountains told of death that awaited those who ventured into the mountains. These Indians, mainly the Pimas, were afraid to go into the Superstitions for fear of Apache reprisal or wrathful gods in the high cliffs. This was sacred land to the Apache and they guarded it with zeal. Only the bravest of strangers or the unknowing dared passage.

The Apaches have a legend about the manner in which they came to be sole guardians of the Superstition Mountains. Many years ago, it is said, there lived at the eastern end of the Superstition Mountains a people called "people of the south" (probably the same Indians as the present day Pinal or Mohave Apache). Their chief was a very proud man who reigned over the village in back of a peak called Wee-Veak-Ah (Middle Rock).

One day the chief decided to take his band of warriors and raid one of the Pima-Maricopa villages that lay west of the Superstitions. He wanted to acquire a few horses and perhaps some food and the sedentary valley farmers were usually easy targets. After he had gathered his warriors and left, the Apache women who remained took the children and set forth to the south side of the mountain to collect some agave and cactus fruit which was ripe at that time.

A day had passed since the braves departed. The women picked all the fruit they could carry and decided to return to the village. On their way they tired and upon reaching a shady and watered spot, decided to rest awhile. After eating some of the fruit and drinking the refreshing water, they became drowsy and were soon fast asleep.

In the meantime a group of Pima-Maricopa warriors, infuriated at the recent pillage of their village, had taken up arms and headed for the Apache camp. In traversing a small valley that led to the settlement, they happened upon the sleeping women and children. The enraged Pimas and Maricopas immediatedly killed all of the sleeping Indians and then proceeded to burn and ransack the nearby dwellings.

At this very time the chief of the Apaches and his band of braves were returning from their recent foray. The leader noticed smoke coming from the settlement and immediatly percieved what had happened. He quickly ordered his men, who numbered about fifteen, to climb up the sides of a steep canyon, through which the marauders had to pass. Here they hid and waited. The Pimas and Maricopas, who numbered 200, slowly entered the canyon where the Apaches waited. At a time when all the men were in the confines of the small and narrow gorge, the Apache chief gave orders to start rolling boulders onto the enemy. The invaders were competely caught off guard and were helpless against the onslaught of rocks and arrows. Even as outnumbered as they were, the Apaches managed to kill all of the Pima and Maricopa forces while suffering no losses themselves. From that day on, no Indian other than the Apache dared venture into the Superstition Mountains for fear of the dread tribe that dwelled there. Thus, in this manner the Apaches came to hold sole right to the range.

The Pimas and Maricopas tell of a similar incident, but their version says that it was not the Apaches who killed their men but an angry god who lived in the cliffs.

Another Pima legend concerning Superstition Mountain tells of the great flood of long ago and how people, trying to escape the rising waters, climbed to the top of the cliffs. They fought each other, vainly trying to reach the safest spot, and mass hysteria prevailed. The Pima god, angered at the behavior of the people, cast his crystal upon the rocks, and in doing so halted the rising waters and turned the people to stone.

In former times, the elder Pimas would take the young men of the tribe to the base of the mountains and show them the many rocky bumps and human-like formations high in the cliffs above. These, they told the youngsters, were the people who were turned to stone. They also called attention to the massive chalky cliff with a crest shaped like a great wave ready to break. This is where the flood waters halted their ascent. Even today, a white limestone ledge can be seen, attesting the height reached by the legendary flood. The old men also pointed to the abundance of sea shells that litter the desert floor, adding veracity to the tale.

In his book ***When the Red Gods Made Man,*** Will R. Robinson states that "It is often said by white men, that in former times, the Pima Indians, as a tribe, never dared venture into the Superstition Mountains, believing if they did that the spell evoked when South Doctor broke his crystal still hung over the place and that they too, would be turned to stone. The continuance of danger of sudden petrification to visiting Indians may have been added by a few of the Pima story tellers to give a spooky touch to the old legend, but in general, this fear was not held by the warriors, and the caution they always maintained in entering the mountain was not an endorsement of this belief but (grew) from the fact that it was an Apache stronghold."

There have been many legends about Superstition Mountain handed down by the Indians who lived near or in its confines. Each has had special, and sometimes religious, meaning. The story of Apache gods inhabiting the mountains played an important part in the legend of the Lost Dutchman; for it was they who were supposed to put a curse on anyone who dared hunt for the Dutchman's gold; and it was to this curse that many unexplained deaths were attributed.

The concept of an Apache "Thunder God," who supposedly abided in the Superstitions, could be traced to the Pima story about the gods who rolled rocks from the cliffs above; and too, the great storms that periodically gathered over the mountains were usually accompanied by "thunderous" roars. So when the "Thunder God" spoke, it was either through the howling of a desert storm or crashing of boulders through lonely canyons.

But the name "Thunder God" came from the white man. There was a tendency to romanticize the Indian legends and make them more palatable for the public. In fact ***Thunder God's Gold*** became the name of a novel, and later, the story was made into a movie that starred Glenn Ford as Jacob Waltz, the Dutchman. The Indians had their own name for the gods but the whites found them too hard to pronounce or remember, thus the evolution of names such as "Thunder God".

Another Pima story told of the people and gods of Superstition Mountain. They gave a brief explanation of this belief to the ***Phoenix Gazette*** in 1893:

> *Montezuma was a great chief and ruler over thousands of souls, the inhabitants of very large cities and populating the extensive*

plains of this country. Fearing that a great calamity was about to befall him and his people, he caused them to assemble on the plains adjacent to the mountain (Superstition) and then, with his magic wand, he caused an opening to be made in the side of the mountain, into which he and his people went. Then the stone gateway was closed and to this day Montezuma and his people dwell within the center of the rugged mountain.

The association of Montezuma with the Superstition Mountains was not new, being held by many Indians and whites alike. There is a persistent belief that the greater part of the Aztec wealth lies buried somewhere in the mountains, and many think the gold Jacob Waltz is said to have found was part of this heralded treasure. More recently, Celeste Maria Jones, who gave much of her life in pursuit of the Lost Dutchman, firmly believed that Weaver's Needle, the rocky pinnacle deep in the Superstitions, is the place where the treasure is hidden. Time and again she had claimed to "see" figures appear from within the Needle, strengthening her conviction that the Needle is hollow and the storeplace of a fabulous treasure. One had only to listen and watch as she sang in the clear desert evening, lifting her strong clear voice to the lofty cliffs above, to be aware of the fact that this indeed was sacred to her.

In 1893, just two years after Waltz's death, the ***Phoenix Republican*** told of an Indian who had brought gold out of the Superstitions. It said that:

The recent discovery of some rich nuggets of pure gold has given new impetus to the mining excitement in the Superstition Mounains.

It goes on to say that the Indian had brought out $150 in color in one chunk but that "superstition kept the Indians from giving information. The fear is based on the tradition that in early days Montezuma enslaved the natives and compelled them to work in the mines."

Indeed, there is another legend that states, upon seeing the Spanish *conquistadores* return to their land for a second time, the Aztecs gathered all of their remaining wealth, consisting of raw gold and gold ornaments, and loading it upon the backs of hundreds of slaves, trekked to the north and away from the Spanish invaders. Traveling for many days, they finally arrived at a mysterious mountain that rose mightily from the desert floor. Here, in the secretiveness of crisscrossing cliffs and canyons, they hid the treasure, burying with it the many slaves who had transported it. A curse was then put on the treasure to protect it from intruders for all time.

Countless legends have arisen from the Indians. The Apaches, Pimas, Maricopas, Aztecs, and others have contributed to the ever growing mystery and enchantment of the mountain called Superstition.

vignette 8

"...the man in Scottsdale was one of her 'angels'..."

Maria Jones was an enigma. On one hand rational, informed, graceful, educated, charming, funny, friendly and conventional. A passerby wouldn't give her a second glance. On the other hand, she could be contradictory, confusing, myopic, extremely unconventional, muddled, hair-brained and unthinking. Her presence in the Superstitions during the 50's and 60's is the stuff of legends and tall

MARIA JONES AND LOUIE
This is one of the few photographs known to have been taken of Maria in the Superstitions.

tales. Who really was she? What did she believe? Why was she there?

Although I wasn't with her long enough to establish a firm rapport, I believed that Maria Jones honestly thought she had the key to finding the lost Superstition gold. Some writers have said that Maria was looking for lost Jesuit gold, and although this may be true, while I was with her she never mentioned it. Nor did I ever see or hear her talk about a map she was supposed to have. My impression was that whatever information she had was handed down to her through family stories and traditions, beginning with her grandfather. She talked about Julia Thomas, the black woman present when Jacob Waltz, the Dutchman, died; and she talked often about Montezuma and how he may have come north with his treasure horde after being run out by the Spanish. To her, I believe, Montezuma's gold and the Lost Dutchman were one and the same.

I don't believe the story that her search for the lost gold was prompted by a tip from a Los Angeles astrologer. Her information, to her at least, was much more reliable than that. She did believe somewhat in these peripheral personalities, but their only influence was that of a confidence or hope builder, nothing more.

Maria had money, not lots of it, but enough to keep a fairly large camp running for a number of years. Where she got it is anybody's guess.

I once made a trip with her and Raymond to the Scottsdale area where we stopped at a house in a well-to-do neighborhood. As I waited in the car, a nicely-dressed, middle-aged, balding man ushered them into his house. In about a half hour, they returned, smiling and in better spirits than before, and now equipped with money to go and buy the supplies we needed. Maria never talked about money or where she got it. I've often wondered if the man in Scottsdale was one of her "angels," and if he was, I marvelled at her ability to convince him to invest in her search for the Superstition gold.

> "...of all the hidden mines of Arizona there is at present the best evidence that the one operated by Dutch Jacob did exist, and that it is an unusually rich claim. The old man in his dying hour made the location so plain to the woman that she has never doubted him."
>
> *The Saturday Evening Review*
> Phoenix, Arizona
> November 17, 1894

The Eternal Search

The decade following the death of Jacob Waltz was the time when the legend of the Lost Dutchman Mine grew at a most frantic pace. These were years in which his contemporaries searched vigorously for the mine, only to find despair and disillusionment—the same story repeated by thousands of hopeful people in the following years. In the 1890's memories were still fresh, and hopes, although crushed again and again, remained high. Yes, these were the times that held the best chances of finding the mine—if it actually existed.

Probably the first person to look for Jacob Waltz's mine after his death was Mrs. J. E. Thomas, the same woman in whose house Waltz had spent his dying days. To her, according to legend, he is said to have given the secret whereabouts of his mine, and although this can be debated, a **very** strong point in its favor is the following, which appeared less than a year after the Dutchman died. The ***Arizona Enterprise***, a Tucson paper, in an article taken from the ***Phoenix Gazette***, reported a rather unusual occurrence when it said:

> Mrs. E. W. Thomas, formerly of the Thomas' ice cream parlors, is now in the Superstition mountains engaged in a work usually deemed strange to the woman's sphere. She is prospecting for a lost mine, to the location of which she believes she holds the key. But somehow, she has failed, after two months work to locate the bonanza, though aided by two men. The story of the mine is founded upon the usual death bed revelations of the ancient miner usual in such cases. There is also a lost cabin connected with it. Its location is supposed to be a short distance back from the western end of the main Superstition mountains.

Less than a year after Jacob Waltz died in her home, Mrs. Thomas, accompanied by two men, was searching the Superstitions for his mine. To say the least, a woman in those mountains was completely out of her element, and knowing next to nothing about mining, prospecting, or the outdoors, her efforts were doomed from the start. But she was there, in the middle of the blazing hot summer, looking, searching, leaving no stone unturned in her quest for Waltz's hidden gold.

It seems highly unlikely that Mrs. Thomas—more at home behind the counter of an ice cream parlor than in the mountains— would venture into some of the most forbidding terrain in the entire West merely for recreational purposes. Conversely, it seems *very* likely that Waltz gave her some type of information relating to his hidden bonanza, and probably in need of money, she decided to take her chances at finding it.

Julia Thomas had come to Phoenix from Colorado City, Texas, where on December 28, 1883, she had "intermarried" Emil W. Thomas. At the time of Waltz's death, she had been a resident of Phoenix for more than 3 years. Her husband, however, wasn't with her for the entirety of the Phoenix years, for their marriage had soured and on March 23, 1890, he "deserted and abandoned" Julia.

As Jacob Waltz lay dying in Julia's home, other concerns lay heavy on her mind also. Just two months prior to the Dutchman's death, Julia appeared in District Court and filed a complaint against her husband, asking that the court grant her a divorce and give her "such other relief as may to the court seem just and equitable." According to Julia's deposition, Emil W. Thomas had "wilfully and without cause" left and abandoned her in March, 1890, and had literally left her holding the bag. But, a determined and perseverant woman she must have been, for even though her husband had left property that was "encumbered with debts to the full value of same," in the 18 months since his desertion she had "claimed and recorded Lot #11 and half of Lot #9, Block #51 in the City of Phoenix as a Homestead, and by her own labor paid off and had cancelled a Mortgage on same for $300.00." Julia also testified that the "fixtures and stock in store left by defendant was of about the value of $300.00 with encumbrances or debts of about $500.00," part of which had been "paid off...with her individual earnings." Deserted by her husband, Julia had put her nose to the grindstone and paid off the mortgage on the community-held real estate and also had made a dent in the debts owed on the store. Her providence and resolve must have impressed the dying Dutchman, no stranger himself to the arduous life.

On October 6, 1891, Julia was again in court. Her errant husband had been located and "duly served with process" in Centralia, Washington. He opted not to appear personally at the divorce proceedings and instead was represented by council. It really didn't matter, for it was Julia's ball game from the start, as the judge ruled in her favor on all counts. The divorce was granted and Julia became a free woman. The land, the store and all personal property were adjudged to be "set apart and turned over to her as her separate and sole property."

Most probably Waltz was under Julia's care and receiving her ministrations during the entirety of the divorce action, from August 22nd when the complaint was filed, to October 6th when the decree was granted. Most assuredly he was with her on the latter date. (One wonders what happened to the provisions of the "lifetime care" agreement Waltz had signed with Andrew Starar some years earlier and why they weren't complied with.) Perhaps it was Waltz's gentle urgings that finally prodded Julia to initiate divorce proceedings, some 18 months after her husband had left her. Perhaps it was also Waltz who had been responsible to some extent in helping her pay off the mortgage on the Homestead and reducing the indebtedness on the store, payment, in kind, for the care he had been receiving, And perhaps it was during this time, too, that Waltz, his health in a steady state of decline, decided to bolster his payments to Julia by dealing her the ace he was holding up his sleeve—he gave her directions to his "mine" in the Superstitions.

This is the point then at which the Lost Dutchman legend began to gather much impetus. With Waltz's last utterances, the seeds were planted, and in the home of Julia Thomas the legend became firmly entrenched. What came afterwards were the thousands of different blossoms that sprouted from its branches. At first only a select few knew about the story of the hidden gold and the knowledge they had was that retained from Waltz's deathbed revelations. Evidently it was not enough, for whatever old Jacob had hidden was never found. Frustrated in her efforts, Mrs. Thomas turned to others. She had taken two men into the mountains on the first trip and they had proved to be little help; now she was forced to seek the assistance of someone who knew that rugged terrain. In doing so, she was not aware of the disturbances she would create in coming years. As she solicited help, the original story given to her by Waltz became more garbled. In her eagerness to direct the new men she had hired, Mrs. Thomas began adding to the original story, in benign hope that she might assist them in finding the mine.

By now the story was gaining momentum. The two men who originally accompanied Julia on the search also turned to others for assistance. One of these men, Reiney Petrasch, was supposed to have been at Julia's house at the time of Waltz's death, so he, too, had the original version of the tale. The number of persons familiar with the death bed revelations steadily increased. Now there was Julia, Reiney and his brother Hermann Petrasch (the latter supposedly accompanied Julia on her first trip to look for the mine), the new men Julia

In the District Court

of the
Second Judicial District, Territory of Arizona,

in and for the County of Maricopa.

Julia Thomas,

                  Plaintiff

    against

Emil W. Thomas,

                  Defendant

Julia Thomas the plaintiff in the above entitled action, complaining of Emil W. Thomas the defendant alleges the following facts as cause of action, to wit: That plaintiff is a resident of the City of Phoenix Maricopa County, Territory of Arizona: That she has resided at said City, County and Territory for more than three years last past: That defendant is a non-resident of said Territory and from information received and belief formed, alleges his residence to be at the Town of Centralia, State of Washington: That plaintiff and defendant are husband and wife: That they intermarried at Colorado City, State of Texas December 28th 1880, and have ever since until March 1890, lived together as husband and wife: That in the month of March 1890, the defendant disregarding his marriage vow, wilfully and without cause left plaintiff with the intention of abandonment, and ever since has deserted and abandoned, and so continues to abandon plaintiff without sufficient cause or reason: That there is nothing issue of their said marriage:

That at the time of said abandonment all of the property held in defendant's name, was Community property, and was encumbered with debts to the full value of the same.

That since said time (March 1890) plaintiff has acquired and recorded Lot 11 [one half] of Lot #9. Block 51 in the City of Phoenix as a Homestead, and by her own labor paid off, and had cancelled a Mortgage on same for $300.00.

That the fixtures and Stock in Store left by defendant was of about the value of $300.00 with encumbrances or debts of about $500.00 part of which has been paid off by plaintiff with her individual earnings.

Wherefore plaintiff prays judgement and decree

1st. Dissolving the bonds of matrimony forever between plaintiff and defendant.

2nd. That the whole of said Community property be set over to plaintiff as her Sole and Separate property. and for such other relief as may to the Court Seem meet and equitable in the premises.

                            H. N. Alexander
                                Atty for Plaintiff

Territory of Arizona
County of Maricopa

Julia Thomas, being duly sworn says: I am the plaintiff in the above entitled action: that the defendant is a non-resident of Arizona Territory. that he resides at Centralia, State of Washington. So affiant is informed and believes.
                          Julia Thomas

Sworn and subscribed to before me this 22nd day of August A.D. 1891. H. N. Alexander

DIVORCE PETITION
Julia Thomas' petition for divorce from Emil Thomas. Note her signature at bottom of second page.

had hired for her second attempt, and the persons in whom Reiney and Hermann had confided.

It was too late to stop the ballooning effect now. From the original handful who knew of Waltz's disclosure, the number grew rapidly. This person told that one, and in turn, he told another—and very few, if any, now knew what they were talking about. The original story, (granting there was one—and it seems likely there was), direct from Waltz's lips, must have been very vague and confusing to begin with, but now it was being altered by second and third-hand speculations. Less than a year after Waltz's death, the information had disseminated to such an extent that the press picked it up—and even offered that the location "is supposed to be a short distance back from the western end of the main Superstition Mountains." It wasn't very long before the "lost mine" had become common knowledge and everybody in Phoenix knew "something" about it. The story was *becoming* a legend and there were already as many versions as the number of people who had heard it. Although nearly three years had passed since Waltz's death, his mine did not have a name. It was referred to as the lost claim in the Superstitions or simply, Dutch Jacob's lost mine. The name "Lost Dutchman" was still to be coined.

As time passed, explicit directions to the mine became public domain, and on November 17, 1894, the ***Saturday Evening Review*** published what it claimed were the details Waltz had given Mrs. Thomas while he was on his deathbed. Mrs. Thomas had visited the editor of the ***Review*** and told him the following: "In a gulch in the Superstition Mountains, the location of which is described by certain landmarks, there is a two-room house in the mouth of a cave on the side of the slope near the gulch. Just across the gulch, about 200 yards, opposite this house in the cave, is a tunnel, well covered up and concealed in the bushes. Here is the mine, the richest in the world, according to Dutch Jacob. Some distance above the tunnel on the side of the mountain is a shaft or incline that is not so steep but one can climb down. This, too, is covered carefully. The shaft goes right down in the midst of the rich gold ledge, where it can be picked off in flakes of almost pure gold."

On two separate occasions Mrs. Thomas had tried to find the mine but failed. She probably now reasoned that there was no harm in making Waltz's story public, and that there was always a possibility someone would discover the bonanza and reward her for the information.

As word of the ***Review*** article spread, the hunt became more frenzied. Soon there was a virtual "trail" of people in and out of the Superstitions, each thinking he could find the mine. One prospector, P. C. Bicknell, found the cave with the house in it, but failed, after persistent effort, to find the mine. The house in the cave was a key landmark in Julia's story (a landmark she had been unable to locate), because the mine was supposed to be "about 200 yards opposite this house in the cave." After Bicknell found the house and cave, the search became concentrated in that area. It seemed like just a matter of time until the mine would be found.

Jim Bark and Huse Ward, who both had ranches near the Superstitions, spent over $1,500 trying to find the spot, but were also unsuccessful. Countless others tried their luck, too, but like those previously, drew a blank. The mine was so near, yet so far away. "But one thing is certain," commented the *Review,* "of all the hidden mines of Arizona there is at present the best evidence that the one operated by Dutch Jacob did exist, and that it is an unusually rich claim. The old man in his dying hour made the location so plain to the woman that she has never doubted him."

By 1896, so much had been said and written about the lost mine of the Superstitions that it finally acquired the name it bears today—the Lost Dutchman Mine. Jim Bark is usually credited with naming the mine after the streams of persons who passed his ranch on their way to look for the "lost" Dutchman's mine.

At the time people were looking for the Lost Dutchman Mine, others were searching the Superstitions for what was known as Doctor Thorne's gold, or more commonly, the Doctor Thorne Mine. Many believed the two "mines" were one and the same, thus adding more speculation to the already garbled Dutchman odyssey.

Dr. Abraham Thorne, while stationed at Fort McDowell, an army outpost near the north edge of the Superstitions, had cured some Apaches of an eye disease that had been plaguing them. As a gesture of gratitude, the Apaches took the blindfolded doctor to a spot in the Superstitions where he was told to fill a small pouch with gold that was lying freely about the ground. It was obviously gold ore that had previously been crushed and hand sorted, leaving the highest grade—a common way of mining in rough, inaccessible country, where transportation and compact loads were at a premium. It may have been gold that was mined and highgraded by the Peraltas, a Mexican family that allegedly worked the Superstition claims some years prior.

After successfully mining the Superstition Mountain country for years, the Peraltas had organized a massive expedition to travel to the mountains and mine all the gold they could before the Treaty of Guadalupe Hidalgo became effective in 1848. This treaty assigned the United States a section of land that had previously belonged to Mexico and of which the Superstitions were a part. The Mexicans had to hurry, for the land they had been mining would soon be United States territory. Hundreds of men, mules, and wagons were sent to the mountains in a last effort, and after extracting as much of the rich ore as possible, the Peraltas sealed the mines and then started on the long, arduous journey south. While still in the Superstitions, the *carreta* was attacked by Apaches. Being unprepared and heavily-laden with gold, the miners were easy victims for the ambushing Indians, and in short turn all were massacred. After tearing open the ore-bearing sacks and scattering the gold, the Apaches confiscated the mules—which were considered a delicacy. Having no use for the gold, they let it lay.

Tales of "massacre gold" became familiar to those who knew the Superstitions and almost equaled in number stories of the Lost Dutchman. The gold Dr. Thorne had picked up was not "in place," that is, not a natural deposit. It showed signs of having been hand-sorted, and in fact, could have been the "massacre gold." Was it also the gold of the Lost Dutchman Mine?

Years later, Dr. Thorne returned to the Superstitions but failed to find the spot the Apaches had taken him to. Although blindfolded by the Indians, Thorne did manage occasional glimpses of the terrain, and one of the landmarks he positively identified was Weaver's Needle. He concluded that the "massacre gold" had to be somewhere "within a circle not more than five miles in diameter and with Weaver's Needle as its center."

At the turn of the century, two prospectors named Silverlock and Malm began digging a series of holes near the north slope of the Superstitions. Although thought to be a bit eccentric for their actions—the north slope was not a likely place to find any type of mineralized deposit—the two "crazy men" soon began shipping considerable amounts of gold ore, the total value of which was to have exceeded $15,000. It was believed that they had accidentally stumbled upon some of the massacre gold, and thinking it to be a rich outcropping, started digging the small potholes in an effort to uncover the mother lode. They never found the source and it is very doubtful that one existed in that particular area. But their discovery posed some important questions. Was their gold part of the Mexican-mined *carreta,* and if so, what happened to the greater portion of it? More important, where were the mines from whence it came? The discovery of Silverlock and Malm documented the now popular belief that indeed the Superstitions were the nesting place of yet to be discovered vast quantities of gold.

There are many versions of how Jacob Waltz was supposed to have found his famous gold mine and most agree that he probably inherited it, in one way or another, from the Mexican Peraltas. Although each story is dotted with grains of fact, none tells—nor ever will tell—the truth about Waltz's adventures prior to his death in 1891. They are but shreds of a much torn and ragged garment.

To arrive at the version which *most closely approximates* the truth, one path becomes evident: the newspapers. Although their accounts may be based on second and third-hand information, the fact remains that this information has not been changed by the passage of time. The stories are there, in black and white, silent witnesses of the past. The newspaper versions of the early legend are those that we today must give most credence. They are the ones told prior to and immediately after Waltz's death; accounts of his contemporaries, as vivid and real today as they were at the turn of the century.

Soon after Jacob Waltz died, an article appeared in **The Prospect,** a newspaper from the booming lumber and mining center of Prescott, Arizona, relating how he happened upon his rich mine. While living in Sonora, Mexico, the story goes, Waltz became intimate with a wealthy Mexican, Don Miguel Peralta. Peralta had discovered and worked many rich gold claims in the Superstition mountains while Arizona was still a part of Mexico. On his last trip to the mines, he and his expedition were attacked by Apaches and many lives were lost. The ferociousness of the previously-peaceful Apaches prompted Don Miguel to abandon the mines and return to his hacienda in Sonora, taking with him all the gold that he could salvage. Not wanting to chance another trip to Arizona territory, Peralta offered the mines to Waltz, giving him a map which showed trails leading to the mines and detailed descriptions of the country. Armed with the map and verbal accounts, Waltz journeyed to Tucson, where he revealed his secret to Jacob Weiser, a countryman whom he persuaded to join in the search for the Peralta properties. Eventually, Waltz and Weiser reached the south side of the Superstition Mountains and proceeded up the first long draw leading to the interior, as per instructions written on the map. They soon found the Mexican's monumented trail which led over the top of the main mountain and down a long ridge, past Sombrero Butte (now known as Weaver's Needle) into a deep, brush-covered canyon. From there the trail meandered through a small boulder-strewn arroyo, eventually terminating at the diggings. As they drew near the mine, the Germans heard someone breaking rock. Upon closer inspection they saw what they thought were three Apaches, pounding and sorting ore. They opened fire and immediately killed the three, but upon examining the bodies discovered that they were not Apaches but peon laborers of Miguel Peralta. The Mexicans had apparently slipped away from the ranch in Sonora and come to the Superstitions to mine gold for themselves. Frightened by their discovery, Waltz and Weiser hurriedly gathered a large quantity of the gold ore left by the victims, and started to make their way out of the mountains.

Meanwhile, the rifle shots had attracted the attention of the

SENTINELS
Eerie "stone guardians" are forever watchful. Formation is just south of the base of Weaver's Needle.

Apaches and before the two men could get away from the mine, they were attacked. In the running skirmish that followed, the two became separated; Weiser, seriously wounded, made his way toward the friendly Pima Indian villages, and Waltz toward the small settlement of Phoenix, each probably believing that the other had been killed.

At that time, Judge John D. Walker, owner of the famous Vekol Mine, was living with the Pimas at one of their villages about eighteen miles west of the present town of Florence. Weiser, after three days of tortuous travel, staggered into his camp, weak and exhausted. Walker did what he could to relieve his pain, but at the end of a week, the German contracted a fever and died. While still rational, he had told Judge Walker about his adventure and gave him the gold he had managed to salvage. Walker was evidently busy with other matters at the time, for he failed to investigate the story, although Weiser gave him full directions on how to find the monumented trail to the mine.

In 1881, Judge Walker repeated the above story in every detail to Tom Weedin, editor of the **Florence Blade**, and promised to take him to look for the mine. However, their proposed trip was never made.

The preceding tale, taken from an early 1890's newspaper, is basic to the legend of the Lost Dutchman. True, many discrepancies exist, but nevertheless, it is probably the one closest to the truth. Waltz was later supposed to have corroborated Weiser's story, and his description and directions were nearly those given to Mr. Walker.

Later published versions of the story add more questions. In one, Waltz supposedly killed Weiser to keep the mine for himself; in another, Weiser was killed by Apaches and buried within sight of the mine. Find his grave and you've found the mine, the story goes. If Weiser did die at John Walker's camp, his grave should have been easy to find. But who at that time was interested? No one. Not until after the death of Waltz did these inconsistencies crop up and by then it was too late. Fact mingled with fiction, history with legend.

It isn't expedient that the exact story of the early Dutchman legend be known, for the manner in which Jacob Waltz found his mine has no direct bearing on subsequent events after his death. These were the crucial years for seekers of the lost mine, and it didn't matter to them whether Waltz found his mine in 1848 or 1862. The important point was that he found it and now it was up for grabs.

We have seen that just a few years after Waltz died, ideas about finding his mine were as many as the people who looked for it. Maps became commonplace and most everyone searching for the mine had access to one. The maps could be traced to many sources, beginning with the Peraltas. Any one of the Mexican miners could have drawn up a map of some type and passed it on to a relative or friend. The recipients could have done likewise. Waltz and Weiser most likely had some kind of map, or at least information that could later be made into one, and so did many Apaches who had observed the mining. Judge Walker supposedly had Weiser's information. Julia Thomas most assuredly had a map of some type etched in her mind, if not given to her by Waltz. Any of those intimate with Julia were in a position to have access to the information. Dr. Thorne had a map, but was it to the Lost Dutchman? The possibilities become rampant and it becomes evident that a chain letter situation was developing—maps became too easily accessible. There were too many maps—from too many sources—for any of them to be of value. Compounding the difficulties, the possibility of fake maps entering the picture made it useless to own one.

The Lost Dutchman gold could be attributed to a great many sources, each having hearsay "evidence" to back it up. There were literally hundreds of people on the trail to the mine, each having the right map, the right information, the right landmarks. Some believed the Lost Dutchman was not a mine at all, but rather the massacre gold Waltz had accidentally found while on a prospecting trip. He claimed it came from a mine in order to throw people off his trail, they said. Others insisted that the gold came from one of the original Peralta mines. Still others believed that his gold was highgraded from one of the early producing gold mines in the Phoenix area. There were other views also and people chose the one that most closely corroborated their information. Armed with their individual solutions, they streamed into the Superstitions—that mine was there somewhere! It was just a matter of time until some lucky soul would fit the parts of the puzzle and claim the riches. But every single one of them forgot a certain day in May of 1887. The day that may have sealed the secret of the Lost Dutchman Mine for posterity.

"...shooting it out."

vignette 9

Ed Piper was a recluse-prospector who made the Superstitions his home. He had a permanent camp about a mile north of Maria's in Boulder Canyon, not far from a year-round water hole. Tall and very thin, sporting a short white goatee, this laconic Kansas emigrant held sway over a rather questionable array of armed characters who called his camp their home. Piper himself contributed to the armed image, always holstering a large gun at his side. He had filed claims on an area around Weaver's Needle and intimated that he was within striking distance of paydirt. Maria Jones had claimed territory near Piper's and the proximity of claims always led to arguments about who was on whose land. While I was with her, the squabble over who could walk this trail and who could climb that hill began to come to a head. There was no love lost between the two and they did their best to avoid seeing each other.

I didn't know Ed Piper very well, only meeting him a few times when I would walk down to his camp to pay a visit. He himself would not tread near the Jones camp. What time I did spend with him was punctuated by cordiality and he always offered me a cup of coffee or something to eat while we chatted. It bothered him that I was associated with the Jones camp and he couldn't see how I fit in. I told him I was just trying to find out more about some of the characters in the Superstitions—himself included—and the Jones camp was as good a place as any to work out of. He disagreed, and echoing Maria's sentiments about him, said the camp was unsafe and that the black woman was crazy. He offered that I stay with him.

Both the Piper and Jones camps had good views of Weaver's Needle and the vicinity where both had filed claims. Caution was the by-word at all times in both camps, and any unnecessary foot travel in the claim areas was avoided. The times I walked down to visit Piper, I made sure that no gun or knife hung from my belt. I thought it better to go completely unarmed, a threat to no one.

One man in Piper's crew was especially proud of his ability with a gun and would commandeer me to watch him fast draw on some tin cans. I didn't know much about fast draws, but he could make his gun clear leather in a split second. I made sure to commend him on his speed and that seemed to make him happy—he seemed like a good person to keep happy—and any good will I could spread between the camps would do no harm. Watching him draw and shoot made me realize that it wouldn't have done me any good to come into the camp armed anyway.

I told Piper that I had come to Arizona to complete my education at one of the universities. From that moment, he called me "college boy," choosing to ignore my real name. He would ask questions about Maria as his stone-faced cohorts sat idly by. What was she doing? Where did she get her money? Who were the people with her? Did I ever see any gold? I answered in generalities and this seemed to satisfy his curiosity—I never gave specifics mainly because I never really knew any, and if by chance I did, figured it was Maria's business and not his. She never did like it when I went down to visit Piper. Each time before I left he would instruct me to tell Maria to stay off his claim and stay away from his water hole (we did go down there to get water a few times). He said he didn't like her trespassing on his property. The mountains are big enough for both of us he would say. There is plenty of room elsewhere for her to do her prospecting.

There was plenty of room in the mountains, but both camps were intent on zeroing in on the terrain around the Needle, and two camps, both heavily armed, both searching the same area and both almost within shouting distance of each other, were destined to eventually settle the problem in one of two ways—by talking it out or by shooting it out. Unfortunately, they chose the latter.

Ed Piper at his Superstition Camp in the early 60's.

(Photo courtesy of Dr. John R. Doran)

/ 67

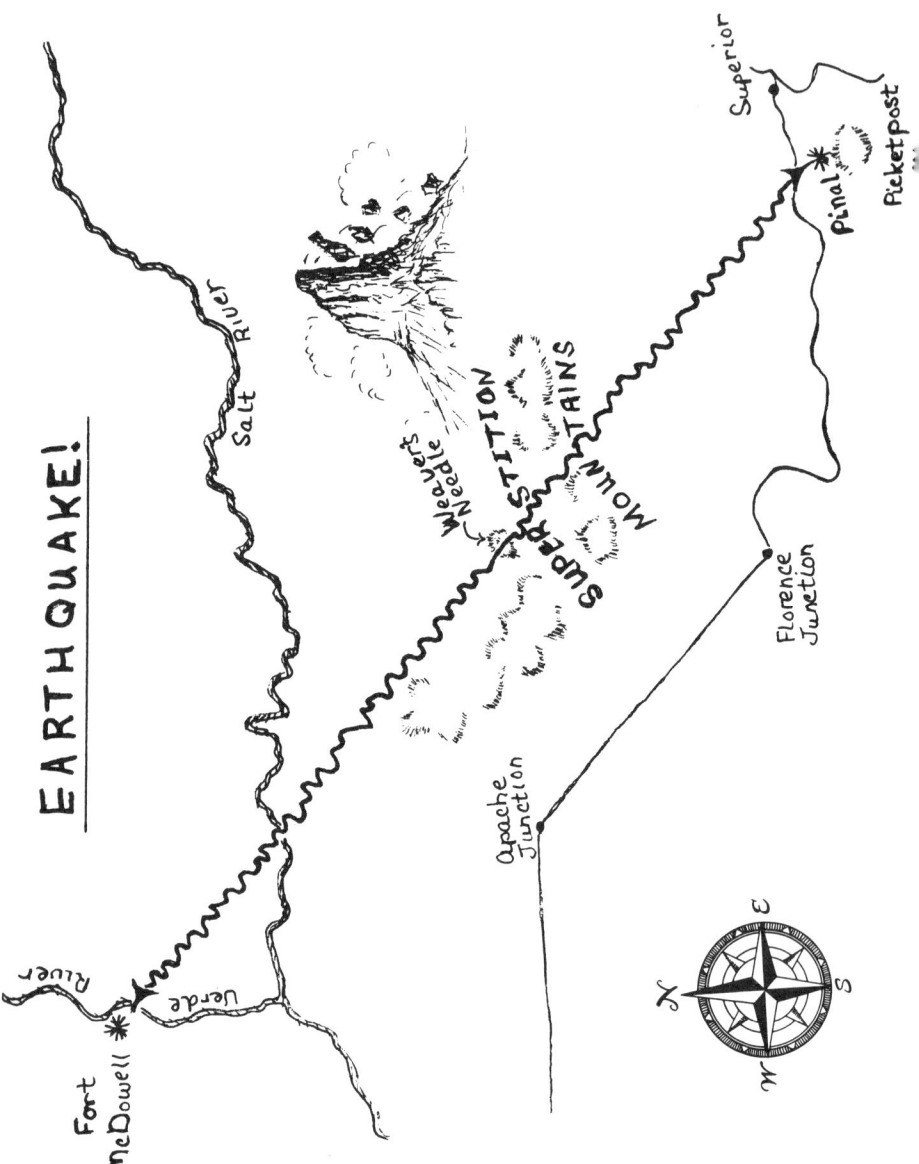

EARTHQUAKE ZONE

Two areas heaviest hit by the May 3, 1887, earthquake were Ft. McDowell and the once-thriving mining town of Pinal, then nestled at the base of Picketpost Mountain. A tremor or quake line drawn between these two points slices directly through the heart of the Superstitions, literally splitting Weaver's Needle. The Superstitions, with their bewildering profusion of high, jagged peaks, precipitous cliffs and steep canyons, suffered extensive damage from the jolt that rocked southern and central Arizona.

9

> Pinal, Arizona, May 3, 1887
> At 2:35 p.m., Florence time, we had quite a sharp shock of earthquake here... Large pieces of rock were detached on all sides of Picket Post mountain, which of course rolled to the bottom, raising a cloud of dust, and for several minutes it ascended about the mountain, giving it the appearance of a live volcano...
> **Arizona Weekly Enterprise,**
> **Florence, Arizona, May 7, 1887**

> Tuesday's earthquake was felt at El Paso, Deming, Tucson, Yuma, Phoenix and McDowell. It seemed the heaviest at the latter place.
> **Arizona Journal-Miner**
> **Prescott, Arizona, May 5, 1887**

> The earthquake was "very heavy at Ft. McDowell, thirty miles east of this city (Phoenix). A large majority of the population rushed into the streets very much frightened by the strange phenomenon, which has never before visited this section."
> **Los Angeles Times, May 4, 1887**

Earthquake!!

The morning of May 3, 1887, bloomed cloudless and hot, typical of the Arizona desert in late spring; the clear skies and already burning sun promised another afternoon of hellish heat, and if anyone or anything moved, it was with slow deliberation. The soldiers at Fort McDowell, a small outpost snuggled against the far north apron of the Superstition Mountains, cursed the encroaching heat. Another day of dust and duties lay ahead. For the soldiers, camp life was a boring routine, and lately it seemed that they were always in camp. Apache raids had become less frequent and the urgency of their first calling had dwindled to an occasional patrol whose purpose was more for exercise than fighting Indians. General Crook himself became somewhat complacent, his obsession with the bothersome Apache much subdued.

The wood wagons had long since left; it had become necessary for the gatherers to go far to obtain necessary fuel for the camp, as the fort had devoured all within its locale. Morning routine was progressing smoothly and chores were clicked off methodically. It looked like another typical desert morning. Beneath their grumblings, civilians and soldiers alike were silently thankful for the promised day of peace.

Nature, however, having the uncanny habit of unleashing one of her freaks when least expected, had other plans. At exactly 10 a.m., the fort was shattered by a thunderous clap from deep within the Superstitions. Canteens and cups rattled and crashed to the ground, wood piles were dislodged, horses bolted, people screamed and babies cried. An earthquake had hit the Superstitions! From Fort McDowell, soldiers watched as huge slabs of rock and debris tumbled down the slopes, taking all that was in their path. Pinnacles shook, depressions were filled, and the hot canyons belched dust in rhythm to the dancing undulating earth. Tons upon tons of rock were strewn over this usually quiet and mysterious land. Familiar landmarks disappeared and topography was altered. The Superstitions were no longer the same; the terrain had undergone a drastic change. It was short, sweet, and very destructive—the Thunder Gods had been angered and revenge was taken. The house had been remodeled. It ended as abruptly as it started and the contrasting silence was tomb-like and oppressive. Not a bird chirped. All life waited to see if the *Dueños* had finished. They had!

The importance of the earthquake of 1887 has been overlooked by almost all who have searched for the Lost Dutchman Mine. This quake, although relatively small in magnitude, did enough damage to make it one of the most important factors in unravelling the Dutchman mystery. According to the legend, Jacob Waltz was to have made his last trip into the mountains around the mid-1880's, a few years prior to the earthquake. All of his descriptions and those of his partner Jacob Weiser concerned the topography existent before the destruction. This is a very important point. It is very probable that Waltz did not return to the Superstitions after the mid-80's. He was now nearing his eighties and the rigors of the prospecting life were replaced by more sedentary occupations.

The validity of any maps or verbal descriptions of the mountains could now be questioned; they dealt with land that had been altered to such an extent that much of it no longer resembled its prior self. The account Waltz gave to Mrs. Thomas seemed lacking as it was, and the advent of the earthquake may have hidden or changed landmarks and clues, rendering them unidentifiable. True, the damage was not such as would make the area completely changed from its previous state, but trails were covered, new ledges exposed, old gullies filled—and perhaps the Dutchman's mine was hidden for all posterity. This is not strictly conjecture, for the epicenter of the earthquake was evidently at or near Weaver's Needle, a landmark central to the Dutchman mystery. Any damage done to the terrain in that vicinity would definitely be detrimental to searchers.

THEME AND FORM
Curious basalt formation in the Superstitions.

Today, evidence of the earthquake and another smaller one that followed a few years later can be seen by anyone willing to spend time packing into the mountains. Weaver's Needle, best known of the landmarks, offers itself as first-hand proof of the turbulence of 1887. Around the base of the Needle, especially to the east, tons and tons of rock, at one time part of the mighty pinnacle, are in evidence. The appearance of the Needle has been changed by the absence of this rock. Environs of the Superstitions are filled with more evidence: canyons are choked with rock and debris which had originally been part of the great cliffs above. Where once there was a water hole, today exists a smooth blending of the surface; a palo verde tree or giant sahuaro cactus, perhaps important trail markers to the mine, now replaced by a blanket of granite boulders; an old cave, across from the entrance to the mine—filled and hidden forever?

Nearly a full century has passed and Mother Nature dispensed her agents well. What the earth tremors didn't accomplish, time did. The wind, sun and rain have done an excellent job of smoothing over the scars on the earth's surface. Now things have settled. The Superstitions have taken on the deceptive appearance of a quiet calm, a smugness that silently laughs at the gold seekers who dare venture there. The winds roll through the canyons, howling with delight, and the storm rains and waters cannot withold their bubbling gurgling glee.

vignette 10
"...using dynamite...on the Needle"

One morning Louie asked me to come with him. He wanted to show me something. About 100 yards from camp was a cave-like opening in the side of a hill. A piece of brown canvas covered the opening, making it hard to see unless you were right on top of it.

Louie pulled the canvas aside. Inside were innumerable sticks of dynamite, blasting caps and fuses, enough, it seemed to me, to move a mountain. This, in fact, is what it was intended for. Louie said that Maria wanted to blast holes in the side of Weaver's Needle to determine if it was hollow inside. If she couldn't find an opening, she intended to make one!

"If the Forest Service catches you doing that, they'll put you away for a long time," I told Louie.

He knew it. Blasting or mining activity of any type on Weaver's Needle itself was prohibited. In spite of this, Maria wanted to go ahead with her plan.

"I don't want anything to do with it," I said, "I don't want to go to jail and I don't want to have any part in damaging a landmark like Weaver's Needle."

I don't think Louie did either, nor Raymond. To my relief, they talked Maria out of using the dynamite, reasoning that there were other ways to determine if any part of the Needle was hollow. She was aware also that as soon as they started packing dynamite toward the Needle, I would begin packing myself out of the mountains. I had made it very plain to Maria that I was completely against using dynamite on the Needle.

The dynamite moratorium didn't last. Not long after I left Maria, I heard or read reports that her crew had been using dynamite in the Superstitions. Whether they actually used any on the Needle itself I never did find out.

10
A Taste of Geology

One of the most controversial geologic areas in Arizona is undoubtedly the Superstition Mountains. They have been the prime target of speculation for nearly a hundred years. Laymen and self-appointed geologists have had a field day conjecturing on the possibilities of gold and other precious metals being present there, but the years have decided nothing. Some agree that mineralization is likely, while others argue that the area is devoid of any precious deposits. Prior to the late 70's there had been no recognized study done on Superstition geology, and even after some work by geologists from Arizona State University and most recently, the U.S. Bureau of Mines, the mountains still remain the target of non-professional edicts.

A rough description of the Superstition Mountains, set down by the Arizona Bureau of Mines in 1959, is on the Geologic Map of Pinal County, Arizona. It shows the whole of Superstition Mountain as an overall composition of dacite. Weaver's Needle is shown as a dike or plug, and directly to the south of it an area of rhyolite outcroppings is noted. Approximately at the Maricopa County line and west of the small town of Goldfield, basalt, locally including tuff and agglomerate, makes an appearance. Along the extreme western and southern aprons, extending to the Peralta Canyon area, outcroppings of andesite exist. This 1959 map until recently was probably the only published professional survey of the Superstitions and is obviously very general in nature.

A rough surface survey of some parts of the interior region was attempted by the author and the findings are here recorded.

The southwestern edge of the main Superstition Mountain, the great cliffs—and the side so popular to photographers—was sampled in three or four different places, directly from the faces of the cliffs. Indications were of some type of volcanic flows of quartz latite or dacite with accompanying hornblende phenocrysts ranging to andesite porphyry. Some small diggings and prospect pits in this same area were sampled and found to be situated in a vein of andesite porphyry, silicified with copper oxide and carbonate minerals, manganese oxides, and possible silver value.

To the east of this region, but still on the southern apron, the Hieroglyphic and Peralta Canyon sections were surface sampled and showed perlite, rhyolite porphyry, and limonite cemented breccia. Again present in the vast bulk of mother rock was quartz latite or dacite with hornblende phenocrysts. To the north, the area around

Weaver's Needle was sampled, with one plug taken directly from the Needle. These showed probable kaolinized monzonite with voids from leached sulfides, some evidencing limonite stains. The many boulders of East Boulder, Little Boulder and West Boulder Canyons were found to be composed of a silicified rhyolite porphyry. Throughout this whole area are many basalt intrusions.

Black Top and Palomino Mountains proved to be somewhat similar to Weaver's Needle in composition. To the west and approaching the First Water area, volcanic tuff, cut by chalcedony veinlets was found to be plentiful. The Parker Pass narrows had abundant rhyolite porphyry breccia, with much basaltic rock again present. The southern edge of Black Mesa showed a tuff-agglomerate composition.

Hackberry Mesa and Box Spring Flats evidenced much chloritized andesite porphyry, brecciated, and ranging to a dacite porphyry with hornblende phenocrysts. The northeastern edge of the Superstitions, near Fish Creek, showed possible sericitized and kaolinized quartz-monzonite. A little more to the west, on Geronimo Head Mountain, volcanic tuff, cut by carbonate veinlets was found.

This is a surficial and by no means complete survey of the geology in the heart of the Lost Dutchman country. Although nothing one way or the other can be proved by it, an interesting question, whose answer is dependent upon such geologic evidence as presented here, can be posed. What is the possibility of the Superstitions being conducive to the "bonanza" type mineral pockets so often associated with them?

In an environment where a great deal of volcanic and hot spring action was at one time present, and where there has been much alteration and silicification (as our survey shows has indeed been the case in the Superstitions), deposits of metal-bearing material *are,* geologically speaking, *very possible.*

Adding to the probability that the mountains could have some extensive metal deposits is the fact that they are adjacent to, and directly in line with a northern extension of the mineralized zone that extends from the copper mining center of Bisbee, all the way north to Jerome. This zone underlies much of the eastern extremities of the Superstitions, and it would be in this area, that is, to the extreme east of Weaver's Needle, that mineralized zones are most likely to occur.

A recently completed (1982) mineral survey conducted in the Superstition Wilderness by the United States Department of Interior, Bureau of Mines, has shed a bit of official light on the geology and mineralogy of the fabled area. The study, entitled "Mineral Investigation of the Superstition Wilderness and Contiguous RARE II Further Planning Areas, Gila, Maricopa, and Pinal Counties, Arizona" is the first government investigation of the area.

A total of 129 samples were taken from various sites throughout the

Superstitions. These included old mine dumps and shafts, prospect walls, exposed veins and veinlets, old diggings and the like. Samples were taken from some of the diggings in the Weaver's Needle area, diggings made by seekers of the Lost Dutchman Mine. Literally hundreds of these holes, pits, shafts and tunnels litter the area within sight of the Needle. The report states that most of these "workings are in unconsolidated, nonstratified surface material or show no evidence of alteration, mineralization, or other geologic indication of mineral deposition."

A search by Bureau of Mines personnel turned up "no patented mining claims and no mineral leases, past or present, within the study area." A patented claim is one in which the owner must prove it would be economically feasible to mine. Federal surveyors and mineral examiners must approve the claim for patent. If approved, the land ownership reverts to the claim holder and he may do with it what he pleases.

Did the survey turn up any appreciable amounts of gold or silver in this legendary home of the Lost Dutchman?

> *Prospect workings containing anomalous amounts of metals were found along the southern border within and adjacent to the Wilderness...*
>
> *Copper, silver, lead, and zinc are found along the southern border in the vicinity of the JF Ranch and Peralta Canyon. Gold and copper occur at the Palmer Mine, which is inside the proposed western addition to the Wilderness.*

The vast majority of the 129 samples analyzed showed only trace or minor amounts of gold or silver. The best gold value was .80 ounces per ton from a sample taken at the old Palmer Mine, just east of Goldfield, off the Apache Highway. Silver values in the JF Ranch area, near the southeast boundary of the Superstition Wilderness, ranged from nil to an impressive 48 ounces per ton. Most of the prospects in this area were covered by current mining claims.

Although most of the samples taken proved to be devoid of precious metal content, the study concludes that: "The occurrences of base and precious metals in prospect workings along the southern border of the Superstition Wilderness in the Peralta and JF areas could be related to deep-seated metal deposits."

Thus, the possibility is indeed very real that someday the mountain range could prove to be the nesting place of economically feasible deposits of ore.

vignette 11

"...one night he would never forget."

Raymond had just dropped Maria, Louie and me off at the Peralta trailhead, the entrance to the Superstitions we always used. Maria had previously arranged to have one of the Apache Junction stables leave a mule there, as we had to haul a number of cans of water in to camp this trip. We were heavily loaded with supplies and everyone carried a backpack while the mule was burdened with the water cans and the remaining gear we couldn't handle. Louie had also brought along an old Italian army rifle and had two bandoliers of shells strapped crisscross, a la Pancho Villa, across his chest.

The sun had already tucked itself behind the great Superstition cliffs to the west and darkness caught us as we wound our way up Peralta trail toward the Fremont Saddle summit. As we stumbled up a particularly steep and rocky portion of the trail, the mule lost its footing and slipped backwards, tumbling over a small ledge and ending up flat on its back about ten feet below, lodged firmly between two boulders. The water cans on each side of the mule were like wedges, holding the beast immobile and unable to do anything except kick and bellow. It was stuck good and in the process of falling had ripped a large gash in one of its rear legs. We dug in the darkness, trying to free the mule from its predicament. Leaking water from the overturned cans, mixed with the dirt and blood, made for a nasty muddy bloody goo to dig in. Sweating, covered with mud and blood, Maria, Louie and I finally managed to work the mule free. It hobbled to its feet. It was hurt bad and there was no way it could continue. We thought it would be best to leave it there with the water and other supplies and Louie and I could return in the morning and take it back down to the trailhead. From there we could get to a ranch phone and notify the stable.

Louie rubbed some type of liniment on its badly cut and bleeding leg and then prepared a mud and leaf poultice and applied it to the gouge. Satisfied that the bleeding was stopped and the mule would be OK until morning, we gathered our gear and resumed the hike towards camp. It was very dark now, and as we trudged over the rim and started the descent to camp, I could see a faint light directly ahead. As we neared, the dying flickers of a campfire became visible, right smack dab in the middle of the trail. And there, near the fire, sitting on his bedroll, was a man.

What a strange place to pick for a camp, I thought as we approached. What's this man doing in the middle of the trail? Then it hit me. If his situation seemed strange to me, how must we look to him? If ever I felt sorry for a person, it was that man camped alone in the Superstitions, sitting there, immersed in his thoughts, enveloped by darkness, his only friend the campfire in front of him.

"How are you doing?" My voice broke the stillness. He didn't reply. I thought about how we must look and what he was thinking as we appeared out of the darkness. Here we were, a black woman, a mustachioed Pancho Villa with a rifle and enough bullets in his bandoliers to start a war, and I, dressed in Marine fatigues, gun at my side, all covered with dried mud and blood and probably smelling worse than we looked.

We introduced ourselves. He didn't say a word, just sat there, frozen by this midnight apparition. I tried to assure him that we meant no harm and as quickly as possible described the mule episode to him, in hopes it would allay what must have been an inordinate fear about our bloody and muddy appearance. The guns I didn't attempt to explain. I'm not sure if any of it sunk in, but eventually he did start to talk, saying he was from Los Angeles and that he came to the Superstitions each year by himself to get away from it all. As we chatted he seemed to relax a bit, even offering us some very old coffee. Although he was talking, I wondered what really was going through his mind and if he believed the story I had told him.

As we prepared to leave, we invited him to stay at our camp which was just a short distance to the north. Not surprisingly, he declined. I thought to myself that this was one night he would never forget. Here would be a story for the grandchildren—I could visualize it growing and being embellished each year until it became a family tradition. And here was one man, I'd bet, who didn't stay the rest of the night at that spot, and who found another mountain range where he could get away from it all.

11
"Lost" Mines

It is likely that the Superstition Mountains have become victim of the "lost" mine syndrome. A curious game that man plays, taking some object or circumstance, adding time, and making it into something other than what it really is. Evidently this is what happened in the case of the Lost Dutchman; but there is no way to be certain. This is what makes the "lost" mine syndrome so lasting.

It would be interesting to question professional people: mining engineers, geologists, and historians, to see what their opinions are concerning the many "lost" mines throughout the Southwest. Probably the majority would remain aloof and refuse comment. But more likely than not their attitudes are best summed up in the following from the booklet, **Mining in Arizona—Its Past—Its Present—Its Future,** published by the Arizona Department of Mineral Resources.

> *Arizona is rich in legends of many "lost" mines.*
>
> *Perhaps 98 percent of the "lost" mines are pure fiction. They exist only in imagination. True, the stories are interesting, especially to newcomers, but they are likewise dangerous. Many lives have been lost searching for these mythical mines, and in addition, the communities are put to extra expense for posses and searching parties.*
>
> *"Don't believe" is sound advice regarding lost mines. Forget the lost mines shown on the "old-map-my-grandfather-bought-from-an old-Spaniard-he-befriended." It simply doesn't exist. The map is probably a fake, regardless of crude lettering, old and soiled paper, or other details which would seem to lend authenticity.*
>
> *The rich ore that is supposed to have been obtained from a "lost" mine—and some were very rich—in all probability was "highgraded" (stolen) from some of the early-day rich mines then working.*
>
> *To a tourist in good health, and accompanied by an experienced prospector, the search for a "lost" mine is a healthful and interesting diversion during the winter months. Such trips always have the possibility of discovering some overlooked mineralized outcrop of promise; a neglected, rather than a "lost" mine.*

The chances of finding *any* "lost" mine, even if it did exist, are infinitesimal at best—and the dangers involved are great compared to possible rewards.

All lost mine stories are *usually* based on some thread of truth. Unfortunately, these threads have been pulled, stretched, woven, and sewn to such degree that they become indistinguishable from the fabric of lies they hold together. Take for example the following account of

COYOTE?
Towering monolithic rock dwarfs a Superstition saguaro.

how one particular "lost mine" came into being:

There was a railroad engineer who had a penchant for collecting rocks. Whenever his train had to stop to take on water or wood he would take advantage of the momentary respite from duties and roam the countryside searching for colorful rock specimens. Over the years he built quite a collection, keeping the location where he found each rock securely fixed in his memory. One day a friend of his, in examining the collection, noticed one rock in particular that seemed to contain much free gold. Upon having the rock assayed, his contention proved correct—the rock *was* exceedingly rich in gold. Questioned as to where he had found the rock, the engineer unhesitatingly named the locale.

Upon returning to this spot, and after extensive searching, they failed to locate the source of the rich rock. Perhaps it was in one of the other spots? And so the hunt started; every one of the places where the engineer said he had picked up some specimens was checked; none contained the gold.

The story spread and others resumed the search where it had been left off. Soon the finding of that one little rock would convert itself into a "lost mine." The source of the gold never was located—indeed, it may have been a piece of "float rock" (rock that was carried by weather or geologic means to a spot far from its origin) that caused the sensation. People, however, thought differently, and the search continued. As time passed, the little rock and its mysterious origin became known as the legend of the "Lost Railroader's Mine." Where in fact a mine never existed, one does now.

There are numerous books about mines of the old West and even encyclopedias which give the location and approximate value of every "lost mine." And, as the Arizona Department of Mineral Resources says: "Perhaps 98 percent of the 'lost' mines are pure fiction. They exist only in imagination."

While the foregoing is undoubtedly true, who can deny we are at least 98 percent richer because these stories exist, and the doubtful two percent keeps the fires of imagination burning. When some stories get old and worn and finally take their place, forgotten in the ash pile of folklore, others, 98 percent lies, will replace them, leaving the necessary two percent to rekindle imaginations. Man creates stories for himself, and when he grows tired, throws them away and creates new ones.

It is a tribute to the legend of the Lost Dutchman Mine that man has not grown tired of it. On the contrary, his interest has increased with passing years. Seemingly not a year slips by without the Dutchman fires being lighted. New twists are added, new maps are found, old manuscripts uncovered, and the "true" story of the mine finally comes to light. And when interest wanes, someone "finds" the Lost Dutchman and the whole process is set in motion again.

vignette 12
"...people on the Needle..."

It was not uncommon for Maria Jones to see people coming in and out of Weaver's Needle. The first time she called my attention to it I was dumbfounded. We were in camp and it was supper time. The Needle, as always around sunset, was illuminated to a brilliant orange-brown and every crack and crevice on its west face was boldly outlined in shadow.

Maria casually asked me if I could see the people coming in and out of the uppermost portion of the spire. I couldn't believe her

question. I looked, straining my eyes for any type of movement—I saw nothing.

"Look! Up there, Bob, near the top. Don't you see them? They're going into a cave. See them, see them!" she shouted frantically.

I looked at Louie. He sat there expressionless, his eyes fixed on the towering precipice. He said nothing—he had been this route before.

"I don't see anything, Maria" I replied.

"They're gone now. It's too late. But they will be back. You'll get a chance to see them again."

I thought just because I didn't see anything, did not mean she didn't. As I found out in subsequent people-sightings, she really believed that she saw people up there and who was I to argue with her? I recalled a time that a friend of mine had tried and tried to point out a walking stick sitting motionless on a tree branch. Try as I might, I couldn't see it. But it was there nevertheless. Perhaps Maria's people were there, too. I hoped for her sake they were.

Maybe she really did see people and maybe they were really going in and out of the Needle. *She* knew they were there, she saw them, she believed in them, she sang songs to them. They were carrying treasure from room to room, from opening to opening, moving it about to confuse searchers of the mine. They were the keepers of the treasure, visible proof that the bonanza was inside the Needle itself.

Maria saw people on the Needle at increasingly frequent intervals. Indeed, by the end of my stay with her, daily sightings were common. I never did see them. Louie was noncommittal, shrugging his shoulders when I asked him if he saw them. I knew he was trying hard to see them, trying hard not to disappoint Maria. He didn't want to crush the dream. He didn't want to say definitely that they weren't there, for saying that would be saying that the treasure wasn't there and the dream would collapse and vanish into the still Superstition air.

12
Death is My Nickname

"Mrs. Laura Branstetter Middaugh, 59-year-old Joplin, Missouri, grandmother, arrived here today en route to Arizona to search for the fabulous Lost Dutchman gold mine," reported the *Arizona Daily Star* on January 24, 1947.

> Leaving her husband, three children and four grandchildren, "who think I'm an old fool," at home, Mrs. Middaugh is hitch-hiking to Arizona in quest of the mine which she says she has located through yellowed maps passed down to her by her great great uncle.
>
> Claiming she is "fit to handle a pick and shovel and smart enough to blow a lode," Mrs. Middaugh told reporters she stopped here (Denver) in search of a grubstake to tide her over during the search.
>
> The doughty grandmother said her great great uncle, Jake Walz, found the Lost Dutchman in the early days of Arizona, and died after carrying out a fortune in gold.
>
> "My husband thinks I'm an old fool, and so do my children," she said, "but I know that the map I have is right, and I plan to find the mine or die trying."

"Or die trying"—words that have haunted searchers of the Lost Dutchman for almost a hundred years! A 59-year-old grandmother, hitch-hiking to Arizona, map in hand, and confident of finding the Dutchman's gold. It wasn't unusual. They are still coming to the Superstitions: hitch-hiking, walking, driving, flying; from far and near. Rich, poor, man, woman—the Lost Dutchman has called them all—the mine belongs to whoever finds it. Prospectors venture into the mountains secure in the knowledge that *they* have the secret to the lost mine, and that *they* have the right map; too many never come back out. These are the ones who chose to "die trying".

Who is responsible for the myriad of unsolved murders in the Superstitions? The answer, as elusive as the killer, has been puzzling law enforcement officers of Pinal and Maricopa counties for years. A great number of persons enter the rage each year: prospecting, hiking, hunting, or merely just curious. Many of these individuals have met death at the hands of unknown assailants. Some have been found with a bullet hole through the head; others were shot and then decapitated; and many just disappeared, never to be heard from again. What's happening in those mountains and who is responsible for the many deaths?

Some blame the Indians; others claim that prospectors, guarding their precious claims, kill those who dare venture too close; the more mystical point to the curse supposedly placed on the treasure hunters by the Apaches. There are hundreds of guesses, but unfortunately they are just that—guesses. Perhaps there is a tiny fragment of truth in each speculation. One thing *is* certain: some person, persons, or thing is responsible, for many of the deaths are not natural. To blame a single person for all the deaths is unreasonable, for too many have been killed over too long a period of time.

Those who hold the Indians at bay for the crimes claim that a curse was put on the Dutchman's mine when the entrance was sealed and it would be guarded forever from the gold-hungry white man. The Indians, after witnessing the Spaniard's greed for the metal of the sun, swore to keep the mine hidden for all time, and anyone who happens on it would be killed. Highly romantic, but possible. If this is so, the Indians must stay very well hidden indeed, for it is a rare day that one is seen in the Superstitions.

Not a single person has been brought to trial on suspicion of being the murderer of Superstition Mountain and it is not likely anyone will be. The puzzle is as baffling a one as ever encountered by law enforcement officers in any place and time. Someone does know the answer, but the mountains forever hold their peace.

Just a half year after Mrs. Middaugh had begun her hike to Arizona, the Superstitions dealt the card of death to another hapless gold seeker. An aura of mystery surrounded the circumstances of the murder of this sixty-two-year-old man, one of the many who had "died trying" to find the Lost Dutchman Mine. He had been flown into the mountains, along with ample supplies, by helicopter. The last persons to see him alive were the pilots. He was not heard of again. On February 21, 1948, over a half year after he had gone into the mountains, **The Arizona Republic** said:

> *Discovery of James A. Cravey, sixty-two-year-old retired photographer, 114 West Polk Street, Phoenix, who disappeared in the rugged Superstition Mountains last June, while seeking the legendary Lost Dutchman Mine, was reported tonight by two Arizona visitors. They are Capt. R. F. Perrin, U. S. Army, retired, and Lt. Commander William F. Clements, of Chicago, guests of the Sunset Trail Ranch, eleven miles east of here. The two men reported finding the skeleton of a man minus the skull, late this afternoon, two and a half miles south of Weaver's Needle, while on an all-day hike in the area. Because of the hour, they did not search for the skull, but brought the man's wallet back to Sunset Trail Ranch. Identification was made through papers in the wallet. Sheriff Cal Boies was notified at Phoenix. Boies said Sheriff Lynn Early of Pinal County will organize a party to pick up the skeleton tomorrow morning. Cravey is the twentieth person known to have lost his life while looking for the fabled lost mine in the Superstitions.*

Cravey was the twentieth *known* person to have lost his life in the mountains, but the actual number was much higher. Many died and were forgotten—unknown and perhaps unwanted. It was a familiar story by now; gold seekers entering the mountains never to be heard from again, paying for the Dutchman's Mine with the most precious commodity—their lives. "Death," indeed was becoming Superstition's nickname.

The following day, Sheriff Early, in the mountains to investigate the murder, found Cravey's skull. It was nowhere near the rest of the skeleton and it appeared that the killer or killers had moved the body and head from their original resting place. One of the most gruesome aspects of the case was the apparent fact that James Cravey had been beheaded! Cravey's body remained in the Superstitions for over a half year without being found and to this day his murderer remains unknown. Who killed him? For what reason? Did he stumble upon the Lost Dutchman Mine? Questions—the answers to which we will never know.

Seventeen years prior to Cravey, another aging man came all the way from Washington, D. C., to look for the lost mine. He had a map, given to him by his son, who, while working as a government cattle inspector at the Texas-Mexico border, had obtained it from a Mexican friend, supposedly a relative of the old Peralta family. He was sure the map was correct and he felt he held the key to the mystery. In June of 1931 he was packed into the Supersitions by two cowhands who worked at a ranch near the southern apron of the mountains. After seeing him comfortably settled in a camp at West Boulder Canyon, the two men bid him good luck and goodbye. They were the last persons

WATERHOLE
Cool pool is a rare occurrence in the Superstitions in summer.

ever to see Adolph Ruth alive. Again, it was almost a half year later that the remains of Ruth, who had spent most of his life as a government worker in the capital city, was found.

"Facing threatening weather and low temperatures, Jeff Adams, deputy sheriff and veteran Arizona cattleman, left yesterday morning for the First Water Ranch at the end of the Superstition Mountains where he will join W. A. 'Tex' Barkley, Mesa cattleman, to begin anew the investigation of the mysterious disappearance of Adolph Ruth, amateur Washington, D. C., prospector. The search was prompted by an archaelogical party sent into the Superstition range ten days ago by the archaelogical commission of the city of Phoenix. The skull is believed to be that of Ruth, who has been missing since last June," so reported *The Arizona Republic* in December, 1931.

> *Adams will be in the mountains four or five days unless adverse weather conditions force him temporarily to abandon the search, J. R. McFadden, Maricopa county sheriff, said yesterday.*
>
> *Adams and Barkley will scour the vicinity in which the archaeological party found the skull in an attempt to locate the skeleton. The belief has been expressed that a wild animal may have dragged the skeleton some distance away.*
>
> *The search by Adams and Barkley, both of whom are acquainted with the treacherous mountain district, probably will get under way early today. Both expected to leave the First Water Ranch yesterday afternoon, and camp last night in the mountains.*
>
> *Ruth, 65 years old, disappeared in his search for the fabulously rich Lost Dutchman mine. He was accompanied into the mountains in June by two ranchers, who returned to his camp several days later with provisions, they told officers, and found he was missing. Failing to locate him in a brief search, they summoned aid.*
>
> *For more than two weeks Maricopa and Pinal County peace officers and ranchers endured torrid summer temperatures of the rock-strewn canyons in seeking the aged man, but found no definite clues.*

The skull was later sent to Washington, D. C., and identified by Ruth's dentist and Dr. Ales Hrdlicka, staff anthropologist of the National Museum. There were two holes in the skull and these were said to have been caused by bullets. But where was the rest of the skeleton?

A few weeks later, in January of 1932, a Phoenix deputy sheriff and a local Superstition Mountain rancher found the remains near West Boulder Canyon, nowhere near the spot where the skull was found. Ruth had been shot twice, authorities conjectured, and then like Cravey, had been decapitated! The murderer or murderers have never been found.

In the pocket of the coat Ruth had been wearing at the time of his death was a little weathered note book. On one of its pages, in Ruth's handwriting, was scribbled: "It lies within an imaginary circle whose diameter is not more than five miles and whose center is marked by the Weaver Needle, about 2,500 ft. high—among a confusion of lesser peaks and mountainous masses of basaltic rock. The first gorge on the south side form the west end of the range," continued the writing, "they found a monumented trail which led them *northward over a lofty ridge,* then downward past Sombrero Butte, into a long canyon running north, and finally to a *tributary* canyon very deep and *rocky* and *densely wooded with a continuous thicket of scrub oak...*"

Below this, Ruth had printed the engimatic words, "Veni, vidi, vici;" and lower yet, "about 200 feet across from cave."

"Veni, vidi, vici"...I came, I saw, I conquered. Did Adolph Ruth find the Lost Dutchman Mine? No one will ever know. Years later, Dr. Erwin Ruth, who originally had given his father the maps of the Lost Dutchman, would again give their reproductions to Glenn Magill, an Oklahoman, in hopes that he might uncover some clues to how his father died. Magill's efforts to find the mine ended as did many before and he came away a frustrated, beaten man.

The list of Superstition's victims reads like a visitor register at a national park. They came from all parts of the United States and many from foreign countries. Their objective was one and the same—find the Dutchman's lost mine. Oregon, California, Colorado, Arizona, New York, Hawaii, Washington D. C., Utah, Australia... the list reads on and on. For some, the date with death was only a short ride away.

Charles Massey came from nearby Tucson, accompanied by a few friends, with plans to do some hunting in the mountains. While in the Superstitions, he became separated from the main party and was never seen alive again. On February 24, 1955, his skeleton was found. Massey had been shot directly between the eyes. A coroner's jury later returned a verdict that his death was the result of a ricocheting bullet! *Was* it accidental or did someone shoot Massey on purpose—and for what reason? Another question that remains unanswered.

A brief year later, the body of Martin G. Zywotho of Brooklyn, New York, was discovered with a bullet hole in the head. Again the coroner's jury ruled the death as accidental or a possible suicide. Under the circumstances, it was as good a ruling as could be expected.

People have wondered how so many unsolved murders can continue, seemingly unnoticed by the law; but the truth of the matter is that the Sheriff's departments do everything possible to solve this lingering mystery. One trip into the interior of the rugged range will convince even the severest critic of the gigantic task faced by lawmen. The mountains cannot be constantly patrolled, and even if it were possible, there is too much area to cover and too many people to watch to make it practical. Perhaps one day someone will solve the riddle, but until then the only way to prevent more killings is to persuade people to stay away from the mountain—and this they will never do. The Dutchman's gold is there, forever luring its seekers to their death.

There were many who were spared the throes of death by the mighty Superstitions. Some have staggered out of the mountains, weakened by severe thirst and hunger, to tell of the unrelenting savageness encountered in those bone-dry, wind-torn canyons. Unwary and overconfident, many have fallen victim to the incorrigible terrain, and those who have escaped did so because of their unyielding determination to live. Many have watched friendships deteriorate and have raised arms against best friends. Some were spared death because instinct told them that the mountain was to blame and the only cure was to get as far away as fast as possible. These were the lucky ones.

In 1934, the Phoenix Chamber of Commerce received a letter from a woman who claimed a narrow escape from death in the Superstitions. *The Arizona Republic* reported the letter in its June 8th edition.

> *A strange chapter was added today to the history of the storied "Lost Dutchman" mine of the Superstition mountains.*
>
> *In a letter to the Phoenix Chamber of Commerce, a person who signed herself Marjorie E. McNulty, Dunsmuir, California, told of being held prisoner in the fabled mine 27 years ago, of a hair-breadth escape and a perilous dash for freedom.*
>
> *"The Lost Dutchman is no myth, but a solid fact," the letter declared.*
>
> *"...The lost mine is not a fable but a real fact—a well-hidden fact.*
>
> *"...I read an article recently in a San Francisco newspaper about that mine ... He (the writer of the article) didn't see the bone yard; he doesn't know of the hundreds who died and were dragged out of sight..."*

Yes ... many—too many—have died in those mountains.

Joseph Kelly was added to the ever-growing list. He came from Dayton, Ohio, and like those before him had benign hopes of discovering the bonanza. He was intrigued by the mystery and thought he had stumbled across the solution. After selling his car, Kelly purchased a gun at a Mesa sporting goods store and then proceeded into the mountains on foot. He left word with Mesa auto court operator Mrs. Grace Boher to call his wife and the police in case he did not return within ten days. Mrs. Boher had to place the calls, for Kelly never returned—nor was he ever heard of again.

"Superstitions Claim Another"—a newspaper heading that became more and more familiar to Arizona residents. The articles began to have a well-known ring: people entering the Superstitions but never coming out. The causes of death were varied, but the most frequent explanation was that it came at the hands of "person or persons unknown."

Bernardo Flores, a 55-year old Coolidge prospector, went into the mountains in search of the lost mine. According to his relatives, "he walked away in the blistering heat of mid-summer and never returned..." His skeleton was found sometime later—it, too, was headless!

"The fabulous Lost Dutchman gold mine, somewhere in Superstition Mountain, off Apache Trail...today threatened another victim," headlined the *Arizona Daily Star* on January 25, 1946. "Missing since Monday without food, shelter, or adequate clothing, is a New York magazine man, Meyer Scuelebtz...a search began yesterday with the assistance of a five-man cowboy posse..."

Unpreparedness...inexperience...unfamiliarity...underestimation—many have laid themselves open to the badge of death. Many of the deaths could have been avoided if only the parties venturing into the Superstitions would take the normal precautions required by any kind of rugged terrain. But, most didn't. They were too anxious and too sure of themselves. No mystery surrounds their deaths—only the belittling of a mighty mountain.

But, there are too many killings that are mysterious...Shot through the head...Decapitated...Who is to blame for these?

In April of 1955, a party of four young hunters entered the range to look for javelinas. During the course of the day, one of the boys became separated, the others thinking he had gone off in another direction to search for the pigs. When he didn't show up at the end of the day, the boys notified the sheriff and a posse was organized to search for the missing hunter. The next day, his body was found, roughly five miles from where he had last been seen. It was lying at the base of a cliff from which he had apparently been shoved; and too— there was a bullet hole between his eyes! Who killed him—and for what possible reason?? There doesn't seem to be any plausible motive for the murder, as the boy came to hunt javelinas, not the lost mine, but he found only death. Another unsolved mystery recorded by Superstition in her book of death!

In 1959, Superstition's rumblings were heard throughout the world. The warm Arizona month of April greeted two friends who had travelled all the way from Hawaii, equipped with a map of dubious origin, eager to look for the Dutchman's gold. They had heard many stories about the legendary lost mine and had finally decided to go to Arizona and look for it themselves. They didn't hunt very long. One night, while he lay in his sleeping bag, Stanley Fernandez was shot and

HILL 3113
Adjacent to the Peralta Canyon Trail, this garbled formation is known simply as "Hill 3113."

killed by his companion and best friend, Benjamin Ferreira. Ferreira later pleaded guilty to a *manslaughter* charge and was sentenced to serve time in the Arizona State Prison at Florence—not far from the Superstitions and the spot where he had killed his friend. After he was released on parole, Ferreira headed back to Hawaii, and after a brief stay, shot his mother-in-law!

The shooting of Robert St. Marie in November of 1959 was another of the senseless killings that have plagued the Superstitions. To be sure, many of the deaths, such as Ruth's and Cravey's, are enigmas; but on the other hand, some are caused by too many people being too anxious to shoot at the slightest provocation. Such was the case of Robert St. Marie, where a childish argument over supposed claim rights led to his death. His killing drew nationwide attention, making the pages of *Newsweek:*

> *Over the years...many a hopeful prospector, armed with crumbling map, a fragment of myth, or simple faith, has explored the Superstitions for the Lost Dutchman Mine. If any of them have found it, they have kept the secret as well as Jacob (Waltz) did.*
>
> *Perhaps the most persistent of the searchers is Ed Piper, a slow-spoken, quick-shooting prospector of 67, who encamped five years ago at the base of Weaver's Needle, a 4,435-foot volcanic spire in the*

shadowy Superstitions...
Three years ago, Piper got some competition, a group led by Celeste Marie Jones, a Negro singer who said she abandoned the concert stage to seek the Lost Dutchman on a tip supplied by a Los Angeles astrologer. Almost immediately, charges of claim-jumping flew between the camps and, since both parties were armed, bullets soon followed. On the grounds that only pistols were necessary for "snakes and things like that and possibly for self-defense," Norman Teason, the justice of the peace at Apache Junction, ordered all rifles confiscated.
Despite the judge's order, Piper subsequently reported, he saw Robert St. Marie, one of Mrs. Jones' crew, approach him, rifle in hand. Piper shot and killed him. "I figured I'd rather stand trial for murder than him," Piper laconically explained. The Arizona authorities, true to the code of the Old West, decided Piper need not stand trial.

In a sworn statement, Piper later testified that Robert St. Marie approached him on the slope of Weaver's Needle with gun in hand. Asked if he was going someplace or looking for something, St. Marie repied, "I will talk with this," pulling a revolver from underneath his coat. Piper immediatedly drew his gun and fired "three times as fast as I could pull the trigger," killing St. Marie. In her testimony, Marie Jones denied the Piper allegations, saying she witnessed Piper and his men open fire without justification on St. Marie. Her account became conflicting when she swore that Piper was still in the Superstitions at six that evening, yelling at the rest of the Jones crew to come out and take their medicine, when in fact he was in Florence at the Sheriff's office. Needless to say, Piper was set free and the killing was termed "self-defense."

"But the search for the Lost Dutchman Mine has been interrupted again," continues *Newsweek*. "Perhaps there is some truth to another part of the Lost Dutchman legend—that the Superstition Mountains are cursed."

Cursed—or just senseless slaughters? Or a combination of both? Less than two weeks after Robert St. Marie met his fate, another man was shot and killed while his wife, petrified, stood and watched—helpless.

On October 23, 1960, some hikers from Phoenix found another headless skeleton in the mountains. Some seventy feet away was the skull... with two holes in it. The remains were later identified as those of Franz Harrier, an Australian exchange student who had been reported missing for some time. What was he doing in the mountains?

Roughly a half year later, on March 21, 1961, the body of Walter J. Mowry of Denver, Colorado, was recovered from its resting place near

Weaver's Needle. A coroner's jury ruled that his death was caused by a gunshot wound inflicted by a person or persons unknown. Seemingly this is as close to the identity of the true murderer as we will ever get.

That same year, the fully-clothed skeleton of Charles Bohen, a Salt Lake City man, was also found. He too had been shot—this time through the heart. Again, the killer was some unknown person, persons, or thing.

The grisly list reads on..."shot through the head," "shot through the heart," "decapitated," "accidental," "person or persons unknown," "ricochet bullet," "pushed off a cliff," "self-defense," "thirst," "fright."

Robert St. Marie was not the only one of Marie Jones' employees to die while in the Superstitions. Vance Bacon, a Phoenix mining engineer, was hired by Mrs. Jones to investigate the east side of Weaver's Needle for possible mineralization. While descending the steep slope, Bacon slipped and fell some 2,000 feet to his death.

In June of 1961, Jay Clapp, a peaceful recluse who lived in a cave in the Superstitions, disappeared.

"The Superstition Mountains, long a breeding ground for fantastic tales, has spawned another mystery story," wrote the ***Tucson Daily Citizen,*** "the disappearence of a man who lived in a mine tunnel and always wore a necktie when he came to town."

> *Pinal and Maricopa county peace officers conducted separate searches last week for Jay Clapp who had been missing for more than two months. Clapp has lived in the Superstitions for 11 years, using an abandoned mine tunnel for a home.*
>
> *Jack Martz, Pinal County deputy sheriff, is thought to be the last person to talk to Clapp. On June 26, Martz gave the 51-year old recluse a ride into the area of First Water, the route Clapp generally took in going to his camp.*
>
> *Clapp's mother, Mrs. Audrey Bailey, arrived here yesterday from Norcatur, Kansas. During the past eleven years, she has sent her son a check every two weeks. Five of her letters lay unclaimed at the post office.*
>
> *Sgt. W. H. Russell, Maricopa County deputy, said that Clapp's camp hasn't been lived in for some time. Found in the tunnel were $42, clothing, a .45 caliber pistol and a camera.*

A pitiful ending was added to this episode when Clapp's mother added, "The last letter I received from Jay was late in June. He loved the Superstition Mountains... he talked as if he wanted to stay there always."

Jay Clapp was granted his wish—he stayed in his beloved mountains and was never seen or heard of again.

As a postscript to the story, Clapp's *headless* skeleton was found

UPRIGHT
Rock formations in East Boulder Canyon. Strange configurations are the rule, not the exception, in the Superstitions.

some three years later, but the cause of death is still "unknown."

The names drag on and on...

"Doc" Burns came all the way from Oregon to search for the Lost Dutchman; instead he found death riding on the wings of a bullet.

Charles Harshbarger and Ron Bley went in—and as far as we know, never came back out.

Guy "Hematite" Frink, an old prospector who had lived in the Superstitions for years, was found dead on one of the many trails in the mountains. He had been shot—and his killer remains "unknown."

An incredible roll call of death! For spells Superstition is quiet and content, but she always erupts in violence when least expected. Although no one can say for certain that the Lost Dutchman is real, they *can* say that the deaths dealt to the mine's seekers have been very real—and permanent. Any person entering the Superstitions, for whichever reasons, should do so only after notifying another party of their plans. If at all possible, take a partner along. Remember—too many have kept silent about their plans and their silence has become permanent. The Superstitions have had a lot of time to practice their dread art, for even as early as 1886, it was "asserted that quite a party of prospectors who entered the range on an organized search for the mother ledge of the famous Silver King were never heard of after."

It has been going on for a hundred odd years, but perhaps the end is finally in sight. The mountains have been quiet for many years and the mysterious and senseless deaths seem a thing of the past.

vignette 13
"...the most forbidden country..."

Nothing would produce a stranger feeling than to sit in the Jones camp as the setting sun played on Weaver's Needle and suddenly hear Maria's voice break out into a strange chant. It didn't happen every day, but when it did it always seemed to catch the camp off guard. She would be in the process of cooking the evening meal or doing some other camp chore when suddenly and without warning she would drop everything, turn slowly toward the majestic burnt orange Needle and begin the chant.

When she sang nothing made sense—it was a mumbo-jumbo foreign to both me and the others in the camp. Perhaps it was something she made up, perhaps something the gypsy had taught her. The closest way to describe it was that it sounded much like a native Jamaican when speaking his own private patois. When she broke into this strange chant, the camp would come to a standstill and all would listen, mesmerized by this black woman's voice. It was like being in an outdoor cathedral; Weaver's Needle the majestic altar; the quiet canyon, the church interior; the air filled with a strange and wonderfully powerful voice coming from the choir. In these moments it was hard to doubt that she really had studied music somewhere and that she indeed had missed her calling, and that the "Celeste" so commonly affixed to her name by the media was not unearned.

Maria would say she was singing to the people in the Needle, the keepers of the treasure. They, she said, liked her singing. On that point I couldn't argue with them. As she sang, it occurred to me how strange, how very strange, that this woman with obvious vocal talent would find herself deep in some of the most forbidden country in the United States, looking for a lost mine or hidden treasure, convinced, entirely convinced, that she had the key to finding it. And how strange, how very strange and really unbelievable that I was here with her.

13

"This is the greatest gold strike in recent years in Arizona."—**Dr. R. A. Aiton, 1920.**

"We don't think we have the right mine, we know we do."—**Glenn Magill, 1965.**

"This time there's no doubt about it."—**Charles Crawford, 1980.**

The Most "Found" Lost Mine in the World

So many people have been in and out of the Superstition Mountains and so many events have taken place there that mere mention of each would take a voluminous work, far beyond the scope of this book. It would also be an impossible task because many persons have passed without leaving a trace of their presence; too, it would be a very boring affair to wade through each of their stories—for many are but stereotypes of others. There are incidents on the other hand, that demand singling out for a closer look. Such is the number of times the Lost Dutchman Mine has been found—so many that you would need more fingers than are on both hands to count them!

The "findings" of the Lost Dutchman constitute a continuous saga of hope and disillusionment, overanxiousness and fraud, and their perpetuation can usually be traced to eagerness of the press for a good story. However, this very fact, in itself, has become a tightly-integrated part of the unfolding story of the mine. The Dutchman's mine has been claimed many times in the past and undoubtedly will be "found" time and again in the future. The "findings" have become as closely associated with the argosy as Waltz himself, and have become an integral part of the legend as we know it today. Arizona newspapers, from the late 1800's to the present, contain many stories of the "findings" of the Lost Dutchman Mine. For the men who claimed to have found the mine the papers offered an excellent springboard into the public's eye.

The first public acknowledgement of the discovery of the Lost Dutchman came right after the turn of the century, as "Charlie Woolf brought in news this week of the discovery of the Lost Dutchman mine, in the Four Peaks country, sixty miles north of Florence. The discoverer is said to be a prospector in the employ of Hi Hooker, the cattle king of the Sulphur Spring Valley. He was not looking for the lost mine," reported the September, 1901, newspaper, "but was simply prospecting around the Four Peaks, and accidentally found the mine. There is an eighty-foot shaft on the new discovery and lying around the mouth of it are several human skeletons and pieces of mining tools. The value of the discovery has not been determined yet."

So, the Dutchman's mine had been found! The necessary conditions seemed to be there all right: skeletons, prospecting equipment, a mining shaft, undetermined value of discovery. But, then the paper goes on to say:

> But the belief that it is the Lost Dutchman mine is, in our opinion erroneous. It is in the wrong locality. All stories of the Lost Dutchman place it on the south side of the Salt River, in a side canyon, somewhere between Weaver's Needle and the south bend of the Salt River, while the Four Peaks country lies on the north side of the Salt River.

The last paragraph, then, added an element that must be present when any lost mine is "found," and that is doubt—the very doubt that will keep the lost mine "lost."

In 1934 the Phoenix Chamber of Commerce was given a hint to the location of the Lost Dutchman. The solution to the puzzle was relatively simple. "Find the gigantic sahuaro cactus, twisted like a creature in a nightmare, and there will be the fabulously rich Lost Dutchman gold mine," advised a letter from Pawtucket, Rhode Island. "Please tell me how many sahuaro cactus, like nightmares, there are in Arizona," the writer incredulously requested. Not knowing much about the description of nightmares, Chamber of Commerce officials replied, somewhat tongue-in-cheek, that the "nightmare variety of sahuaro cactus is unknown to Arizona." With such good substantial clues like the above, it's a wonder the Lost Dutchman hasn't been found more times than it has!

Yet another simple directive was given to seekers of the hidden mine a brief six months after the above give-away. An unnamed prospector, writing to the *Arizona Republic*, said that when "looking from the cave and sighting along this high cliff you can see one, the most northern, of three pinnacle rocks. This pinnacle has been sculptured to represent a German soldier and is about three quarters of a mile from the cave. While staying in the cave," the letter continued, "the Dutchman was fascinated by this pinnacle rock and conceived the idea of using the rock to mark the location of his mine. The statue consists of cap, face, eyes, ears, nose, arms and cape. The face is so carved that no matter from what direction you look at the statue, it is always facing you. If you figure out the place this soldier is looking, you have solved the mystery of the lost mine."

Twisted cacti, sculptured rocks, old yellowed maps, Spanish mining symbols, Indian petroglyphs, faint trails, float gold, caves, rock houses, grandfather's stories, Weaver's Needle, the light of a full moon, the rays of the noontime sun, imagination... countless clues, puzzles, riddles—all leading to the finding of the Lost Dutchman. Every year more clues will be added and new evidence presented. Along with these, the Lost Dutchman will invariably be "found"— over and over again—as the years pass.

A few years ago, one of the strangest persons ever to venture upon the trail of the Lost Dutchman Mine showed up in Phoenix. His proposed method for finding the mine was even stranger. Phoenicians thought they had heard everything when it came to finding the lost mine, but even oldtimers were dumbfounded now. This had to be the last straw! The party involved was going to try to locate the Lost Dutchman Mine by using one of the most unusual and rare talents a human being can possess—extra-sensory perception!!

The amazing part about this caper was that it was on the up and up, and the man whose talents were to be used had been acclaimed by scientists, police departments, and the most stringent critics, to definitely possess the powers of ESP. His name was Peter Hurkos, and his credentials were a backlog of solving strange cases that had baffled conventional detection. His fame was already world-wide and his sincerity and dedication to the truth could not be questioned. His extra-sensory perception was a matter of record.

Peter Hurkos and associates came to Phoenix amidst little fanfare and publicity. It was dutifully announced that he was going to try to locate the Lost Dutchman by using his fabulous gift of ESP. Phoenix waited, it had seen too many people look for the mine and all produced the same results—nothing. Little was heard about the progress Hurkos was making until a somewhat belated report appeared to the effect that he had found the Lost Dutchman and that it was very rich indeed. The report also noted that the mine was not in the locale popularly attributed it. There was no fanfare, no fuss, and not much more could be learned of the episode. The statement was made, and after a brief flurry of interest—and denials by everyone not involved—was passed off as another eccentric gamble.

But, there was something different about this "find"—it slipped by without the usual rituals being performed. Everything seemed to happen too quickly and with a minimum of publicity. This was not typical as far as the Dutchman's Mine was concerned.

Frank Edwards, reporting this event in his book *Strange People,* says of Peter Hurkos: "Today he lives in Milwaukee, happily married and with a substantial income assured, for Hurkos used his weird talents to locate for himself and a small group of associates the fabulous Lost Dutchman gold mine, in Arizona, one of the legendary treasures of the West."

Did Hurkos actually find the Lost Dutchman? Was he, with his fantastic powers of extra sensory perception, able to fathom the century-old mystery? No one knows for sure—only Hurkos can answer that one. Many said he didn't, for after Hurkos came more claims that the Dutchman had been found. And they will continue to come, for the mine will forever, and never, be found.

Another of the stranger "findings" of the Lost Dutchman had a reverse effect on the man who claimed it. Instead of making him rich, it cost him $5,000! This weird situation developed in 1936 when a man stumbled out of the Superstitions with a sack full of pure gold, claiming that it came from Waltz's lost mine. The gold was very fine and pure—*too* pure in fact—and it aroused the suspicions of authorities. The man claimed it had come from the Lost Dutchman Mine; the police investigators disagreed and an inquiry followed. On November 13th a Denver paper said that:

> Calm, precise legal terminology has written an end to the reported rediscovery of the mythical Lost Dutchman mine in Arizona.
>
> Judge J. Foster Symes of Federal District Court listened to the testimony and decided that the bits of gold supposed to have had their origin in the mine "where there are shovelsful of nuggets" were in reality, manufactured from gold intended to fill teeth.
>
> The judge ordered the gold, $5,000 worth confiscated upon the recommendation of the United States Mint authorities.

Upon final analysis, the gold turned out to be stolen dental gold that had been hidden in the Superstitions, and the man who happened upon it, later billing the "find" as the Lost Dutchman Mine, was now $5,000 poorer.

In January, 1935, sheriff's officers in Phoenix were puzzled and somewhat apprehensive over a note which floated down the Salt River in a bottle and was retrieved near Phoenix. "Lost up Salt River. Broken leg. I've found the Lost Dutchman's mine, I think. Come up and I'll reward you. By Blue Point. Please come quick. Jake Lee."

It is doubtful that Jake Lee's Lost Dutchman mine was ever found and also doubtful that many people ventured into the Superstitions to help the man with the broken leg. Poor Jake!

In any situation where a person is offered a chance to radically improve his economic well-being in a short period of time, there exists an element in society only too willing to help—and help themselves in the process. There have been and always will be the sharpies, frauds, hustlers and promoters; they are attracted to the fast money like a bee is to honey. They are the leeches who grab on to a good thing—or if a good thing doesn't exist, they create one. Then they sit back and watch the gullible take the bait.

In a "quick money" deal, the hustler has a tremendous advantage, mainly because he is dealing with human nature. He caters to the greed in man, and once it is set in motion, lets nature run her course. The person who was fleeced usually begs to be taken in the first place. They bait the hook with their own money, swallow it, let the promoters reel them in, dislodge the money, dispose of the carcass, and beat a fast retreat. Meanwhile, the victim sits silently, ashamed to admit he has been bilked out of his hard-earned cash. The history of the Lost Dutchman Mine is clouded by schemes which have attempted to fool the public. Some, indeed, have been very successful. Let's go back to the year 1920.

The most "found" lost mine in the world has been rediscovered! This time by a mining corporation—and this time something was in the wind. But the public, elated at the discovery of their beloved mine, didn't smell the rat. Twenty years had passed since the last find, but the "findings" were to become more frequent with passage of time.

Phoenix residents who picked up the *Arizona Gazette* on August 20, 1920, were delighted to read the following article, and to many it seemed like the chance of a lifetime.

> Word that ore assaying $408.40 to the ton has been uncovered in the famous "Lost Dutchman" mine, situated within 60 miles of Phoenix and 4 miles southeast of Fish Creek hill, reached this city Friday, arousing much interest in mining circles.
>
> This report was verified by Dr. R. A. Aiton, secretary and treasurer of the company which is engaged in developing the "Lost Dutchman" property, who spoke with the greatest enthusiasm of the latest news from the mine.
>
> The sample which the assay just made shows runs $408.40 per ton was taken from the face of the drift on the 100 foot level, he said, and was not "selected" ore in any sense. The gold is carried in a rhyolite-bachial quartz, and is practically a free milling proposition with a vast quantity already in sight.

Great Gold Strike

> "This is the greatest gold strike in recent years in Arizona," said Dr. Aiton. "The 'Lost Dutchman' mine, which was lost in reality for more than forty years, recently was rediscovered and now is in an active state of development. There is an immense ledge uncovered, and enough high grade ore to keep the mine in operation for countless years."
>
> Operations will be pushed swiftly, said Dr. Aiton, and in the near future the company will build its own mill at a convenient site, there being plenty of water available for every need. It is predicted that when the Lost Dutchman begins shipping, the first-class ore will run

far richer than the sample just assayed.
The importance of the strike to the city and section, as well as to the state at large, is pointed out by men prominent in mining circles. The effect of the discovery, and of the subsequent development of the property is bound to stimulate business greatly, it was declared.
No doubt was felt that the company would push its operations without waiting for the improvement of transportation facilities, as the ore is rich enough to make its being packed out by burro-train a paying proposition until a road can be constructed to the mine and mill.

It looked good—too good. However, closer inspection of this "publicity release" reveals some flaws that lie hidden between the lines. Phrases like "vast quantity already in sight." "greatest gold strike in recent years," "enough high grade ore to keep the mine in operation for countless years," "immense ledge," "near future," "rich ore"—all smack of a phony come-on.

But the public, undaunted, wanted to believe that the Lost Dutchman mine had been found. Readers of the **Gazette**, and subsequent newspapers to carry the story, couldn't wait to get their hands on part of the action.

Using the newspaper's undying thirst for a good story as a lair, the new owners of the Lost Dutchman neatly laid their bait. Everything looked too good, and as the story said, "The importance of the strike to this city and section, as well as the state at large, is pointed out by men prominent in mining circles." Who were the prominent men? Remember also that business was supposed to be stimulated greatly because of the discovery and its subsequent development. To the undiscerning eye it looked on the up and up. Prominent men, newspaper publicity, business stimulation—people wondered, is it possible to invest in the mine??

It so happened that the Lost Dutchman Mining Corporation, as the promoters chose to call themselves, had set up its offices right in downtown Phoenix—and of all the convenient things, was offering stock to the public! On the company's application for stock was the statement "Our ninety day credit plan may interest you." These boys were years ahead of their time!

What happened? The subsequent development of the mine never materialized, and the mine itself proved to be a gross exaggeration. However, the money that lined the pockets of the officals of the Lost Dutchman Mining Corporation was no exaggeration. It was very, very real.

And so it goes. The passing years are dotted with such cases; they have come and gone, and once again Old Superstition has the last laugh.

Some forty-five years after the demise of the Lost Dutchman Mining Corporation, another corporation was formed with an amazingly similiar name: The Lost Dutchman Exploration Company! This time it was for real—the Lost Dutchman had been found. Glenn Magill, a private investigator from Oklahoma City, and a group of associates had laid claim to Waltz's gold mine. They had incorporated and were now in the process of developing the mine. Publicity was rampant. Heady statements by Magill flashed across newswires. "We don't think we have the right mine, we know we do. I believe the mine is even richer than anyone can imagine. The Dutchman said it had the wealth of many mines. I don't believe he fully understood how wealthy it really was."

A Tucson radio station, learning of the fabulous discovery, received permission from Magill to broadcast reports, *a la* tape recorder, direct from the site of the new-found fabulous gold mine. Adding more credence to the "discovery" was a statement by Sidney Brinkerhoff of the Arizona Historical Society in Tucson, to the effect that Magill's claim to the mine certainly "matches more closely more aspects of the legend than any before recorded. The significance of this discovery," continued Brinkerhoff, "is in a close parallel with descriptions of the mine left to us in years past. If in the weeks ahead, they are to hit pay dirt, it will be because they have gone at this project with a conscientious and scientific approach. Certainly they've put a lot of hard work into the project. And in the end they probably will come closer to the truth than any group in the past."

Brinkerhoff, whose knowledge of the Lost Dutchman Mine was probably limited to what he had read in Sims Ely's book or one of Barney Barnard's pamphlets, was careful to couch his phrasing so that the statement was not an out-and-out confirmation of Magill's claims. However, it did have the effect of lending peripheral academic sanction to the Oklahoman's efforts.

But, here again, something wasn't quite right.

Brinkerhoff had said that "If in the weeks ahead, they are to hit pay dirt..." But Magill already claimed he had pay dirt. He had confirmed press releases relating to his finding the Lost Dutchman—and discovery of the Lost Dutchman Mine must be synonymous with pay dirt—and had even talked of the gold that was taken out. But where was the gold? The Lost Dutchman was supposed to be a "gold" mine, wasn't it?

A bit more critical of the "find" than the national news outlets was the ***Apache Sentinel,*** the local Apache Junction newspaper.

> *A location claim that pinpoints the exact site of what the claimants say is the Lost Dutchman gold mine was filed Thursday afternoon in*

the Pinal county recorder's office, Florence.

The filing came as a direct follow-up to the nationwide publicity given on a claim that the fabulously rich gold mine had been located in the Superstition Mountains...

Despite the filing of the claim, most Apache Junction residents and visitors most closely familiar with the gold searches over the years remained skeptical.

The skeptics cited the absence of any material evidence to corroborate the announced find. Rather, all that anyone had to go on was the earlier pronouncement of Oklahoma City private investigator Glenn Magill.

The paper went on to say that Magill had found the mine by using maps that had originally belonged to Dr. Erwin Ruth, whose father, Adolph Ruth was slain while looking for the Lost Dutchman. Magill's claim of removing nuggets of almost pure gold from the mine was doubted, however, because "He did not show any of the nuggets. Nor did he show the pictures he said were taken of the mine and of the men breaking a seal placed on the mine by Apache Indians."

The mine Magill was supposed to have found was beginning to look more and more like Mark Twain's "hole in the ground." Indeed, that's what it turned out to be!

In the final analysis, this recent Lost Dutchman "find" never did turn up any gold, as was the pattern of the many before it. A few pieces of quartz, flecked with gold—not an uncommon find for even an amateur prospector—was the extent of the fabulous bonanza. Now that the heat was on, Magill began to squirm, finally admitting that his previous statements were perhaps a bit premature. Maybe the mine is just a little more to the east, or the west; perhaps if he looked...

On June 7, 1967, the Oklahoma Securities Commission filed a "Cease and Desist" order against the Lost Dutchman Exploration Company. The officers had been selling stock in the company and failed to first register it with the Securities Commission. Magill quit the Superstitions and went back to his Oklahoma.

In 1932 Thomas Wiggins caused what was probably one of the West's final all-out gold rushes when he showed up in the small mining town of Superior, just a few miles east of the Superstition range, with handfuls of gold nuggets as big as marbles! The discovery was made, Wiggins said, in the Superstition Mountains, about six miles from Superior. Upon assaying, Wiggins' gold ore proved to average about $12,000 per ton, making it fantastically rich.

"It is believed to be the famous Lost Dutchman mine," heralded **The Arizona Republic**, "about which innumerable legends have been woven and which untold persons have sought for years. Gold stampedes were in progress to the site...Residents from all over the mining district were flocking in large numbers to the little gulley on whose edge may be seen a ledge of gold. So rich is the ore that when broken it is still held together by ribbons of almost pure raw gold."

Things got so bad that armed guards were soon hired to watch over the claim. Hundreds of people quickly filed claims on "every bit of land for miles around the rugged Superstitions" and the county recorder's office was flooded with inquiries. It was gold rush days all over again.

News of the strike spread rapidly and soon the small town was bustling with activity. Scores came from nearby Globe and Miami and many traversed the fifty-odd miles from Phoenix, all wanting to see the extent of the rumored gold ledge.

"Already every room and space in the town is occupied," said the **Republic**. "Tents are being put up and one enterprising trader from 'outside' has his entire stock of goods piled in the streets..." And this was 1932!

But, it wasn't the Lost Dutchman, just a very rich pocket of ore that played out in a few days. As quickly as it started, the rush subsided, and the Superstitions enveloped themselves in silence again. But, who could now doubt that there *was* gold in the mountains?

In 1949 a brassiere salesman from Los Angeles, who had no prior knowledge of prospecting or the Superstitions, laid claim to the Lost Dutchman mine. He said he found it as easy as one, two, three—all he needed was a map and a guide. On July 31 of that year ***The Denver Post*** reported his discovery:

> The century-old riddle of the location of the famed Lost Dutchman Mine in Arizona's Superstition Mountain country was believed solved Saturday.
>
> Answer to the whereabouts of the lost mine for which scores of persons met violent death over the years was indicated in mining claims filed by Henry H. Bruderlin, 35, of West Los Angeles.
>
> Bruderlin's claims are to an area thirty-five miles east of Phoenix in the hills surrounding the myth-ridden Superstition Mountains. He thinks he has the original bonanza, last worked by the mad Dutchman, Jacob Walz (sic), in the middle of the last century.
>
> Bruderlin's luck began when he was given a map kept for years in the family records of an old Mexican family of Ray, Arizona. Next he enlisted the aid of Jess Mullins, 75-year-old prospector who has spent the last sixty years combing the Superstitions for gold.
>
> Following the faded lines on the map, Bruderlin and Mullins were led about four miles out of the Superstition's proper into the foothills where they picked up an ancient Spanish trail.
>
> After that, Bruderlin said, it was merely a matter of following Spanish mining symbols hacked into saguaro cacti and chiseled onto scattered boulders.
>
> The trail led to a hill set apart. Seven shafts were there—seven

abandoned caved-in Spanish mining pits, Bruderlin told International News Service.

"It all seemed so simple—lifting the veil of a century. But it wasn't without its thrills. We came to the top of the hill, and there they were—seven wonderful tumbled-in-shafts. I forgot for a moment that this was the twentieth century."

Bruderlin also forgot that these particular old shafts he "discovered" were known about long before he even thought about the Superstitions. To be sure, they were evidence of the past mining activity by the Spaniards, but they weren't the Lost Dutchman—just seven very old diggings without the slightest trace of gold. Bruderlin soon became disillusioned with his lot, and like the many before him who claimed to have found the evasive mine, faded into oblivion.

And so it goes...

In April of 1975, Robert "Crazy Jake" Jacob held a press conference to announce to the world that he and another prospector had found the Lost Dutchman Mine. It was in the Superstitions, and, like legend has it, was within view of Weaver's Needle. Jacob said that the entrance to the mine was covered with "only 18 inches of dirt and some two by fours and rock. Anyone could have opened it and closed it within 20 minutes time." It seemed incredible that, after more than a hundred years and countless numbers of searchers, the mine was so easily accessible. But, people have a tendency to overlook the obvious and perhaps this had been the granddaddy of all the overlooks. In an AP story, Jacob said "I knew exactly where we were going, but we didn't know positively it was the Dutchman until Friday morning." The article goes on to add the one ingredient so common to all of the Dutchman "finds": "Neither Jacob nor Burkett presented any gold at the press conference to substantiate their claim of finding the mine..."

And so it goes...

Two of the most recent Dutchman "finds" were announced within weeks of each other. In January 1980, Charles Kentworthy, head of a treasure hunting corporation, announced (to the media, of course) that he and his group had found a number of gold and silver "glory holes" in the Superstitions. Kentworthy said they used aerial photography, the infamous "Peralta Stone Maps," a proton magnetometer and even an optical sensing device to locate the bonanza. However, Kentworthy was a bit more cautious than some of his predecessors about claiming outright that it was the Lost Dutchman. When asked by a reporter if his find was the Lost Dutchman Mine, he replied "Who can tell? Nobody ever put a sign on it." (Kentworthy must not have seen "The Better Half" cartoon reprinted here!)

"Okay! . . . Two jerks mean you're stuck, three jerks, go for help—and one big jerk is what you are for going in there in the first place!"

(Reprinted by permission of Bob Barnes)

The Phoenix Gazette quoted a Tonto National Forest spokesman in Phoenix on the Kentworthy claim: "There have been many claims and none of them have been borne out." A classic understatement if there ever was one. What made the Kentworthy claim even more incredulous was that the very same "Peralta Stone Maps" used to locate his find were said to have been used to help find gold near Santa Fe, New Mexico, over 350 miles away!

A few weeks after the Kentworthy "discovery," another man announced that he had found the Lost Dutchman Mine. He added a new twist to the myriad of Dutchman "finds" by claiming he found the mine by "dowsing." In February, 1980, Charles Crawford told a ***Phoenix Gazette*** reporter that "This time there's no doubt about it." He found the mine in La Barge Canyon, exactly where the old maps he had said it would be. It was just a simple matter of interpreting the maps correctly.

The mine had been sealed by tons of rock that was blasted from the cliffs above. This is in contrast to "Crazy Jake's" mine which had been covered with only 18 inches of dirt and rock. An old tailings dump nearby assayed at 3 ounces of gold per ton, and by Crawford's estimate, that pile alone would contain 20 million dollars worth of gold—and he hadn't even got to the good stuff yet!

What happened to this find? Contrary to most Lost Dutchman claimants, Crawford was still around in 1982, giving slide presentations on his "find" at $5 per person, and conducting horseback tours to his "mine" where participants were allowed to pan or dig for gold—all rather unusual activities for someone who just two short years ago had claimed to have found the "richest mine in the world."

As this book is being readied for printing, another man claims he has found the Dutchman's illusive mine. Like others before him, he said he used the now infamous "Peralta Stone Maps" to lead him to the discovery.

In an ***Arizona Republic*** story dated Feb. 6, 1983, Michael Bilbry, a transplanted Los Angeles computer technician, said he "found the mine site at a point that is touched by the shadow of a nearby peak when the sun is in the right location." Bilbry, true to the form of others who previously had "found" the mine, was already promising to spend the money that the mine would make him. He was going to buy, according to the ***Republic***, an F-15 fighter plane! "I'll get one used," Bilbry boasted, "for $25 million."

Bilbry said he figured out that the "Peralta Stones" had originated with the early Spanish Jesuits (a claim made by others, also) who had worked the mine site and left the stone maps as clues to other Jesuits who would follow, a contention some present-day Jesuit historians find ludicrous, without foundation and completely untenable.

What will happen to this latest Dutchman "find"? Only time will tell.

There are more "finds" to tell about but the endings are invariably the same. The Lost Dutchman Mine will never be found, not really. It will remain hidden, lost in the rubble of historical discards, its identity forever masked by the veil of passing time. For those who claimed to have found it and for those who will do so in the future, there will be others to refute them. This is how it has been and most likely how it will be. It is, if you think about it, also how it should be.

"Two hundred feet across from the cave," said Ruth . . .

"On Bluff Springs Mountain," said Ely . . .

"Near Four Peaks," said Charlie Woolf . . .

"In LaBarge Canyon, exactly where the maps say it is," said Charles Crawford . . .

"Within view of Weaver's Needle," said "Crazy Jake" . . .

The many wonders of the legend become manifold as each day etches itself into the past—gathering bits of lore from the years, as it grows and propels itself to the very pinnacle of western adventure. As the legend grows, one can hear—if he listens closely—Waltz's warning to the hundreds who would dare hunt for his mine—"It is very hard to find."

The foregoing theme was so persistent that it became part of the legend itself. Searchers of the Dutchman's mine, some sincere, some looking for attention, laid claim to the lost mine. These "findings" in themselves have become inexorably intwined in the legend and have been added as another chapter. Truly, the Lost Dutchman is *the most "found" lost mine in the world!*

vignette 14
"Somebody...was...following me."

    Raymond had just arrived in camp to pick up Maria and Louie, and another man who had spent a day with us, and they were now on their way back to Phoenix. I had the luxury of having the camp to myself for a whole day. Because it was cooler than usual, I thought I'd take advantage of the brief respite from the heat and do some exploring on my own. On previous excursions I had spotted some interesting outcroppings near Bluff Springs Mountain and thought this a good time to take a close look at them.

    I spent much of that day roaming around looking for possible mineralization. As the afternoon wore on, I worked my way in a southerly direction, eventually ending up near Fremont Saddle, close to Peralta trail. The sun was dropping fast, so I decided to head back to camp. I picked up the trail heading north—it would take me almost directly to camp. As I wound my way down the twisting trail, immersed in my thoughts and the splendor of the Superstition landscape, I caught a brief glimpse of something moving, high above me in the cliffs to my right. I kept walking, head straight ahead but eyes twisted to the right, straining to catch the movement again. There! Dashing across an opening, there it was again, visible for a split second. I couldn't make out what it was, a mountain lion or smaller cat perhaps. I kept walking, a little faster this time, keeping my eyes peeled on the hills to my right, feeling a bit desolate and vulnerable. Then I caught a good look at my tracking companion high above me—it was a man!

    It's an odd feeling to be in the Superstitions and know that your every move is being watched. Somebody, for some reason, was discreetly following me. I wondered why. How long had he been there? Did he see me digging around Bluff Springs? What did he want? Uneasiness gnawed at me as I saw the man another time or two, always to my right, always high above me. Somewhere south of Weaver's Needle I lost track of him, or more accurately, he gave up his efforts at tracking me. Perhaps he had decided that I was no threat or perhaps his curiosity had been quenched. I felt better.

    When I arrived back at the camp, I immediately noticed a large sharpened stick, much like a crude spear, sticking in the ground in front of my tent. It wasn't there when I left that morning. Someone had been here. I checked out the camp, but nothing seemed disturbed, everything was in place except for the new addition—the crude spear in front of my tent. What, if anything, did it mean? Why did someone put it there, just to show they had been in the camp? It didn't make sense. Was there some connection between the man I had spotted following me and the stick near my tent? I mulled over these things as I tried to find sleep that evening, wishing that one of the crew was there. Someone had been in the camp and someone had been following me. Why? My uneasiness was eventually overcome by fatigue as I dropped off to sleep, still pondering the day's events.

The next day, after the crew returned, I asked Louie about the man who had followed me. He said it was not unusual to have someone watching you while in the Superstitions, it had happened before to them a number of times. But the sharpened stick they had no answer for. It would remain a mystery.

14
The "Peralta" Stone Maps

About 25 years ago a new twist was added to the Lost Dutchman legend—the "Peralta" Stone Maps. These stone maps or tablets were supposedly found by a vacationing Oregon man and his wife who had pulled off the highway that skirts the southern apron of the Superstitions to get a better look at the mountains. The stones were discovered a short distance from the parked car, not far from the hamlet of Florence Junction.

The alleged discovery of these stone maps has renewed interest in the search for the Lost Dutchman, and, as we have just seen, a number of persons have claimed that the stones led them to the bonanza, bonanzas, however, which to date have been bereft of gold. As the names "Pedro" and "Miguel" appear on the stones it wasn't long before enthusiasts added "Peralta," alluding to the Pedro and Miguel Peralta of the early Dutchman legend. Thus the name "Peralta Stones" was coined. The stone maps have added yet another facet to the already inflated Dutchman story and in the short time since their "discovery" have become as controversial as the legend itself.

Not many years after their discovery, the "Peralta Stones" became the focal point of litigation in Phoenix, the court eventually ruling that the stones be handed over to a non-profit, mineral-oriented organization. As a result of this ruling, the stones became the property of the A. L. Flagg Foundation, a non-profit Phoenix-based organization that loans its considerable mineral collection and resources to various museums and groups. The "Peralta Stones" are currently on permanent loan to, and displayed at, the Mesa Museum in Mesa, Arizona.

Are the "Peralta Stones" for real, or are they just another of the many fake maps that supposedly lead to the Dutchman's lost gold? Charles W. Polzer, S. J., noted Jesuit ethnohistorian at the Southwestern Mission Research Center of the Arizona State Museum, and widely recognized authority on the Spanish Colonial period in Mexico and Arizona, has serious reservations.

"First of all," Rev. Polzer says, "the stone surface is milled which one would hardly expect from a frontier person in the rough. The incisions were made by tools of very late and more probably early 20th century design."

Of Stone #1 he says: "#1 is not as finely milled as #2. The 'Jesuit' is a poor rendition of a Salem witch. Both crosses on that figure are 20th century design. The lettering is typical of 20th century sanserif type fonts; the ciphers employed are of industrial design and both together are anything but colonial Spanish. The vocabulary is incorrect; but no

"THIS PATH IS DANGEROUS"
On Stone #1, front side, the Spanish text reads: "This path is dangerous. I go 18 places. Look for the map. Look for the heart." Loosely translated, the Spanish could also read: "Search the map. Search the heart." The "secret" tabs have been placed there by the A. L. Flagg Foundation.

one would consider bad orthography as proof or disproof of anything. The reverse side of #1 with the horse is weird because the draught horse shown was not exactly the kind of horse known to Spanish or Mexican immigrants in the area. As a matter of fact the horse design on the rock can be found in children's drawing books."

As Rev. Polzer notes there are a number of misspelled words on the front and back sides of Stone #1. "Vereda" is misspelled as "Bereda" and "Voy" is spelled "Boy." The word "peligrosa" is written as "peligroza." Spanish language experts who have looked at this writing agree that the substitution of the "B" for "V" in Spanish is a very common and acceptable "mistake." Spanish doesn't recognize the difference in pronounciation between "B" and "V." They also note that it is very common to misspell the word "peligrosa" as "peligroza," the "s" and "z" having almost identical sounds.

Misspelling the word "caballo" (horse) as "cobollo" is puzzling, however. Language analysts agree that it would be highly unusual for even someone with a cursory knowledge of Spanish to substitute "a" for "o"—especially in such a common word as "caballo." The Spanish word for heart, "corazon," is also incorrect, the "r" being left out. Deliberate misspellings? Clues to be followed in interpreting the stones? Or just a clumsy attempt at a fraudulent scheme?

"NORTH OF THE RIVER"
On Stone #1, reverse side, to the right of the horse, the stone reads: "I pasture north of the river." To the left, the inscription reads: "The horse of Santa Fe" or "The horse of the Holy Faith." A number of misspelled words appear on both sides of Stone #1, the most notable being the common word caballo (horse) misspelled as "cobollo."

Going to Stone #2 Rev. Polzer says: "#2 Map with its finely milled face and neat incisions depicts some familiarity with brands and branding tools. Some of the ripple lines are reminiscent of the incisions made on stones for milling often seen in rural Mexican ranches. The crude letters depicting 'DON' on the opposite side are totally inconsistent with known Spanish letter formation and are totally consistent with English type fonts. And if the suggestion was intended that 'Don' Pedro Peralta or 'Don' Miguel Peralta was indicated by this usage, one must realize that 'Don' in Spanish is never self appropriated—it is always the title of respect by someone **of another.** Hence it would identify nothing. It's pure poppycock."

DOES X MARK THE SPOT?
On Stone #2, front side, arrow points to wavy line with small, evenly-placed holes along it. This line, when followed to its terminus, contains 18 such holes, corresponding perhaps to "18 places" referred to on Stone #1.

THE "DON" PERALTAS?
The word "DON" can barely be seen on Stone #2, reverse side.

Moving on to Stone #3, Rev. Polzer notes: "#3 Map is about the same as #2 in the quality of the face and the nature of the incisions. But more absurd is the heart that is strictly of northern European or Anglo character. Spaniards never depicted the idea of a heart with this kind of geometry. The ciphers were also done in a post industrial age—they're not even good for 1847 if that was intended as a date. The cross on the opposite side of #3 stone is 19th century in design and had carryovers into the 20th; it's a question of what kind of exposure to design the maker had. Also the dagger or knife as depicted is a very poor portrayal of the Bowie knife and at best maintains proportions of 20th century variants of that weapon frequently found after World War I. There are even elements in the design that indicate a knife used as a short bayonet in close combat."

Did the "Peralta" Stones originate with the 19th century Mexican Peraltas of Lost Dutchman fame or were they carved by early Spanish Jesuits, as a number of people contend—or are they a product of 20th century scheming and fraud? The foregoing comments, Rev. Polzer believes, are satisfactory to indicate that the "Peralta" Stones are fraudulent. He notes in a letter however, that "I must reserve an extensive commentary because some items not only are indicators of fraud but point to the perpetrators."

Are the "Peralta" Stones genuine? Individuals must judge for themselves.

HEART OF THE PUZZLE
On Stone #3, front side, the heart is in place and the wavy line with holes continues onto the heart to its end. A knife or dagger is seen at the far left.

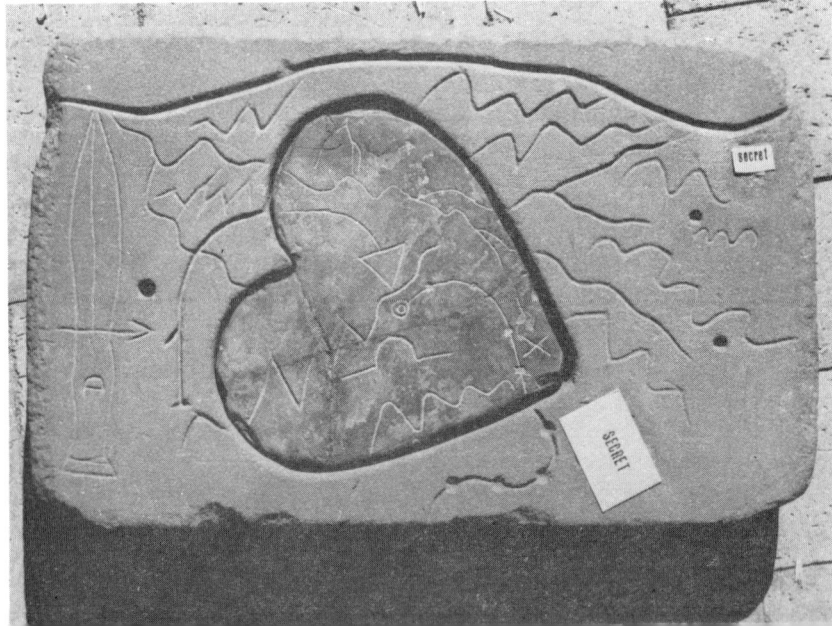

A CROSS TO BEAR
This neatly-carved cross appears on reverse side of Stone #3.

THE HEART ITSELF
The heart, made from a different type of rock than the tablets, reveals cracks where it has been broken and glued back together.

/ 115

SIX FIGURES
Reverse side of heart reveals six oblong or rectangular figures.

HEART REMOVED
Stone #3 with heart removed exposes a neatly-chiseled "1847." A date? or some other significant number? Notice the optical illusion in this photograph. The heart appears at times to be sitting on top of the stone. In actuality, the photograph shows the hole the heart fits into.

HALF-BURIED HEART
With a bit of imagination, Weaver's Needle appears like a heart, half-buried in the ground. Treasure hunters have correlated this heart-like appearance of the Needle with the heart in the "Peralta Stones." The command on the stones, "Look for the heart," has led many to believe that Weaver's Needle is the "heart" referred to.

(Photos in this chapter (pages 111-116) taken by the author with permission, courtesy of the A. L. Flagg Foundation and Mesa Museum, Mesa, Arizona.)

vignette 15
"...someone was shooting into our camp."

One day in early April, an incident occurred that permanently altered my association with the Jones crew and the Superstitions. We had spent the whole day tramping around the area between Black Top and Bluff Springs Mountains and I was dog tired. I couldn't wait to get back to camp. Maria had cooked a large meal, and after eating, all I wanted to do was to go to my tent and rest. I took my shoes off and lay there, half dozing, half looking through the open flap at the sun-drenched hills above me. Suddenly, I heard a thud on the ground immediately outside the tent. A split second later came the distant echoing of a rifle shot. It didn't dawn on me at first, but then the thud and the rifle echo came again and I realized that someone was shooting into our camp. Maria began to scream, "Come on, Bob, come on!"

I charged out of the tent barefooted. Maria and Louie were already running, heading for the safety of some boulders to the west of camp. I wasted no time following them. Four more shots and four more thuds. The best we could tell they were coming from the vicinity of the hills directly to our east, high above us. In camp we would have been sitting ducks. But, to make matters worse, in our haste to get out of camp, we had run in the direction of the dynamite cache and now were just a few yards from it. If one of those shots should hit the dynamite, it would be all over for the three of us. Somebody up there was after us or wanted to scare us—if it was scaring they were after, they succeeded. I prayed that the shooting would stop. It did.

We hid behind the boulders until dusk, then, under the safety of darkness, returned cautiously to camp. I put out what remained of the campfire, afraid that the glowing coals might make an ideal target for our tormenter. I had no idea who might have done the shooting or for what reason; Maria and Louie on the other hand, were convinced that it was that "crazy man" Piper.

That night as I lay in my tent, sleep was not my companion. I thought long and hard about this adventure I had embarked on and realized that the time had come for me to call it quits. Those first two shots slamming into the ground just a few feet from my tent were all I needed to make up my mind. I thought about Maria and the crew. They just seemed to be treading water, more obsessed with the burgeoning affair with Piper than with finding the mine. I didn't like what was happening between the two rivals and figured it was only a matter of time until someone got hurt. I didn't want it to be me. I decided that I wanted nothing more to do with this errant crew.

The next morning, after eating breakfast with one eye on my eggs and one eye on the hills above, I asked Maria if we needed supplies and volunteered to go to town to get them. She said Raymond was bringing some engineer in that day and I could go back with him. He could then drive me back to the trailhead. This sounded perfect.

A few hours later, Raymond and the engineer arrived. They asked me to stay and point out some outcroppings we had discovered, but I said it was already late and it would be better if we waited until tomorrow to get a fresh start. And too, considering the events of the prior evening, we all needed a day off. Everyone agreed that this was a good idea and settled into camp, the engineer looking a bit nervous after hearing about the shooting.

Raymond and I began the trek up Boulder Canyon and on to his car which he had left at the Peralta trailhead. The plan was to buy supplies in Mesa (Raymond had brought some stuff with him, but I convinced Maria we needed some other things), and then he would bring me back to the trail where I could lug them into camp. As we neared Mesa, I told Raymond that I was not going back, this was it, I was quitting.

"I'm not surprised," he said. "After those gun shots I really don't blame you."

Raymond always seemed to me the realistic and common sense carrier of the bunch. He didn't try to talk me out of it and even offered to take me to my friend's place in Tempe. I declined his kindness, preferring to break my association with the crew right then and there. I could hitchhike the short distance into Tempe.

"What about your gear, don't you want it?" he asked

I showed him my knapsack and patted my gun and holster.

"Tell Louie he can have anything that I may have left."

"Does Maria owe you any money?" he inquired.

"Forget it. She can pay me back when she finds the treasure, and I really hope she finds it, Raymond, I really do."

I felt a tremendous pang of empathy for this man. I liked him, I hoped for his sake, they would find something somewhere, but deep inside I knew better. I only hoped that somewhere he would find his Lost Dutchman, whatever it might be.

We pulled off the road. Raymond offered me his hand and warmly touched my shoulder with the other.

"Goodbye, Bob, and good luck! You know, of course, Maria will be very disappointed," he added.

"I know, Raymond, I know. My best to everyone and no hard feelings."

"No hard feelings."

I got out of the car and closed the door. Raymond pulled a long U-turn and headed back east, toward the Superstitions. As the Chevy worked its way through the afternoon heat, he waved. I felt a tremendous sense of relief. It had been quite an experience, much more than I had bargained for, and I had enjoyed much of it. But I had grown increasingly uneasy as I could see events develop around me. The shots into camp the night before had catalyzed my decision, one which I had made none too soon.

The Chevy grew smaller, a diminishing speck of blue, and finally disappeared into the mirage of reflected heat waves. It was the end of the story—I never saw Maria Jones or a member of her crew again.

15
What Happened to All the Gold?

A few years after Jacob Waltz died, a number of different versions of the story of his mine spread, and even though details varied, the stories revolved about a central theme—Waltz *had* a mine or cache of some type in the Superstition Mountains and he had taken gold from it at times. The amount of gold he was supposed to have taken was valued between several hundred and several thousand dollars. No-where can we find anything to indicate he had taken out greater amounts. Remember, this was back in the 1890's, and these were the accounts of people who were contemporaries of his. Old Jacob was supposed to have said his mine was the richest in all the world, but seldom, if ever, did he confirm such braggadocio. However, it cannot be denied that on occasion he was seen in Phoenix, Florence, or Prescott carrying small amounts of gold nuggets. But, so were many other people. Waltz's hermit-like personality and his probable re-luctance to talk about the source of his gold may have given over to speculation about whether or not he had more gold than he cared to talk about. It was only natural, in an environment of prospecting and mining activity, that people would wonder about his gold, no matter what quantity, and where it came from. That he never bothered to register a legal mining claim in the Superstitions also may have led to some questions.

To say that Waltz was a rich man, and that he spent his gold lavishly, is foolish. Every source we have attesting to his financial status, even word of mouth, pictures a man of meagre means. Remember, when he signed that legal document in 1878, he was broke, and twelve years later at the time of his death was still living in an adobe hut, located in the "poor" section of Phoenix. If he had a substantial amount of gold, what did he do with it?

In those days business transactions could be paid for with gold in its natural state, but these, however, were usually small, consisting of such things as buying groceries, supplies and feed. Anyone who had access to large amounts of raw gold usually shipped it to one of the major mints and in return was paid in currency. This is what Waltz likely would have done had he access to substantial amounts of gold. Phoenix was small, and it is doubtful that any of the gold handlers would have kept enough cash on hand to buy large quantities.

One writer on the Lost Dutchman claims that between the years of 1881 and 1889, Waltz shipped $254,000 worth of gold to the Sacramento Mint! This is a fantastic sum to accumulate in such a short time. Why then did he die a pauper? That writer, now deceased, said he personally traveled to Washington, D. C., and examined the archives of the United States Treasury to obtain the above information. Indeed, if this information were correct, it would be a very strong point in favor of Waltz's supposed claim of fantastic riches. Surely he must have had a bonanza somewhere to account for that large sum of money.

As with many facets of the Lost Dutchman tale, however, closer inspection of these "facts" reveals a few inconsistencies.

Number One: A letter from Eva Adams, Director of the Mint, United States Treasury Department in Washington, states that no such records as the above writer claimed to have examined in his "lifetime" search for truth about Waltz, ever existed.

Number Two: The claim that the National Archives in Washington, D. C., held the records of the Sacramento Mint is also false—there never was a *United States Mint* established in Sacramento.

Number Three: Waltz is said to have shipped his gold to the San Francisco Mint also. James K. Otsuki, Chief of the Reference Service Branch of the Federal Records Center, General Services Administration, the government agency that keeps the old San Francisco Mint records, writes: "We have been unable to locate any reference to Jacob Waltz, Woltz, Walz, Walzer, or Waltzer in the deposit ledgers, registers, and journals of the San Francisco Mint."

Number Four: Two other United States Mints were in existence at that time: a Branch Mint in Carson City, Nevada, founded in 1870 and shut down in 1933; and the main Mint at Denver. Records of the Denver Mint are in the custody of the Federal Records Center in Denver, and those of the Carson City Mint are in the National Archives in Washington. Inquiries to these places brought like replies: no records of Jacob Waltz having ever shipped gold ore to either of the Mints. So, where do we go from here? Another possibility exists, but it has doubtful merit.

Many privately-operated establishments for the coinage of gold were in existence in the West during the very early days prior to and just after the opening of the United States Mint at San Francisco in 1854. These mints were in no way connected with the national minting system. Did Waltz ship his gold to one of these mints? It isn't likely because he didn't even arrive in Arizona until 1862, six years after the San Francisco Mint had been in operation. By this time most of the private gold-coining establishments were closed and the larger government mints had taken over. Waltz, at this early date, if in possession of any substantial quantities of gold, had no choice but to ship it to San Francisco—which he didn't. And, too, it is highly unlikely that the mine was discovered so soon after his arrival in Arizona. There have been claims that the Wells Fargo Stage records were checked and they verified that Waltz shipped gold, via their services, to the San Francisco Mint. Irene Simpson, director of the History Room of the Wells Fargo Bank in San Francisco, says that whatever records existed covering the years that Waltz lived in Arizona, were lost in the great San Francisco fire of 1906. So this claim, too, is false. There were never any records in the first place *to* check.

Where does this lead us? A blind alley? Why all the talk about fantastic riches? Why even $254,000? Granted, it sounds good, but it is a lie. Every possible record that could be checked refutes any statement or claim about Waltz's wealth, and there is *no* evidence of any type to show that he had large sums of money or gold at any point during his life. There is every reason to believe, on the other hand, that the great wealth attributed to Waltz and his mine was the product of the imaginations of the men who followed—or even of Waltz himself.

But, what if Waltz was the miser type? He could have hidden and saved the gold he found, content to get by with the bare necessities, receiving more satisfaction from hoarding his gold than from spending it—a virtual Silas Marner. And, too, why couldn't he have used an alias when shipping his gold to the Mints? It would have been easy to use a different name each time he made a shipment. These are, of course, possibilities; and if you so choose, you may believe them to be the actual case. There is no way to disprove it. Just let the imagination roam, for without it, the legend of the Lost Dutchman would never have been created.

Since the death of Waltz, thousands of people have hunted for his mine, and as we have already seen, many claimed to have found it. No one, at anytime, has shown any gold. But no one has given up hope— that mine is still there...somewhere. They hunt today as they have in the past; ignoring the odds stacked against them; ignoring all evidence to the contrary; ignoring the dangers; ignoring the disappointment; and most importantly, ignoring the death that lurks in the mystery-ridden mountains.

16
A Look at Tomorrow

All legends are flexible and can adapt themselves to suit the needs of the writers, artists, their audiences, the general population, and the times. Such has been the case with the Lost Dutchman. The legendary home of the mine, the rugged Superstition Mountains, are now changing at a rate far more rapid than the story of the Lost Dutchman itself. Things are not the same there anymore, and the mountains, not wanting to be at odds with the legend, are being glossed over. It is both good and bad, this renovation process that is taking place. It will have the effect of lessening the number of murders which can only be good—and it will let more and more people come to this once-forbidden land to see for themselves the haunts of the Dutchman. It is bad in that it will take much of the mystery, enchantment and solitude out of the mountains, for along with the increasing number of people comes the onslaught of civilization and "progress." The final outcome is hidden somewhere in the future.

The Superstitions are part of the great Tonto National Forest and fall within an area designated as the Superstition Wilderness Area, a 125,000-acre tract of land set aside by the Forest Service to be preserved in its natural state. No roads, logging operations, or transportation (mechanized or motorized) are permitted. All of the important Lost Dutchman landmarks fall within this wilderness area. Under permit, cattle graze over the 60,000 gross acre Superstition Allotment. Additionally, three other grazing allotments also fall within the Superstition Wilderness Area.

As of June 10, 1965, by Public Land Order No. 3684, a 320-acre section of land, with Weaver's Needle as its approximate center, was set aside as a public landmark and recreation area—not even mining claims are allowed. In addition, 13 other areas in the Superstition Wilderness have been formally withdrawn by the Forest Service from any type of mineral exploration. All of these, excluding the ones near Dons' Camp and Weaver's Needle, were withdrawn to protect water sources thereabouts. In all, a total of approximately 710 acres have been affected by the formal mineral withdrawal and no claims or mining activities are permitted within their confines. Unknowing prospectors, now probing around the Needle and other withdrawn areas, may be unaware that even if they should happen upon a mineralized zone, claiming it or mining it would not be permitted. And most importantly, by virtue of the Wilderness Act, there will be no more mining claims of any type permitted in the *entire* 125,000-acre

SIGNPOST
A Forest Service sign greets prospective Superstition hikers at the First Water Trailhead. Persons using main Forest Service trailheads must register during the cooler "busy" season.

Superstition Wilderness Area after January 1, 1984. However, *valid* existing claims made prior to that date will be honored.

Where once only the hardy and adventurous dared pass are now well-travelled Forest Service trails which beckon the television trapped. Trail names are boldly etched on sturdy sign posts. Dutchman's Trail, that way...Peralta Canyon Trail, this way...the signs indicate. Trails are so good that, if permissible, one could ride a motorcycle or a small car right to the base of Weaver's Needle without much problem. Boy Scout troops, western lore enthusiasts, rock hounds, hunters, well-heeled dudes astride horses and various and sundry persons now enter this once-dreaded range.

The number of persons now hiking into the Superstition Wilderness is mind-boggling. In the first four months of 1982, the Forest Service logged 14,000 people entering the mountains by way of the Peralta Trailhead and another 8,000 through First Water—22,000 people using the wilderness area in four months' time! These figures do not include the numerous persons who also gained access through the less popular routes of Hieroglyphics Canyon, Boulder, J. F., Roger's Trough, Reavis and Tortilla Trailheads. All told, over 50,000 people were recorded by Forest Service personnel as using the Superstition

Wilderness during the approximate duration of the "busy period," an astounding average of over 400 persons per day. Indeed, the western half of the Superstition Wilderness Area is being used and abused by a number of people far exceeding what anyone had ever anticipated. The Forest Service, in order to protect this once pristine terrain, may have to limit the number of persons using the area sometime in the near future. This is especially true of weekends, when it is virtually inpossible to walk any distance at all without seeing other people—a situation that makes rangers shudder. The whole concept of a wilderness area, in the case of the Superstitions, will have to be re-examined, for it has become, as one gentleman recently remarked to me, an "urban" wilderness area.

The great legends spawned in the Superstitions have helped keep a vital part of the Southwest alive and real, but if the current trend continues unabated, one cannot help but hear the shouts of the barker as he lures the curious:

"Step right up, ladies and gentlemen, and get your tickets for the ride into the mysterious Superstition Mountains. Yes, folks, one of the last remaining wilderness spots in America today. Now you, too, can see the same sights and travel the same country as did the old Dutchman many years ago. Only ten dollars the round trip, children half price.

"Don't worry, folks, it's perfectly safe, riding these new rail trams is just like sleeping in your own bed. And it's air-conditioned, too! Hurry, folks, visit one of the least smog-infested areas in the entire Southwest. And remember, ladies and gentlemen, the ten dollars includes a light lunch at the world-famous Needle Restaurant, high atop Weaver's Needle, in the very heart of the Superstitions. Get your tickets now, folks. the next tram leaves in twenty minutes. Step right up, ladies and gentlemen, and ..." ...And the old and weary Dutchman winces and turns over once more—perhaps for the last time.

EPILOGUE

In November of 1959, a number of months after I walked away from the Jones camp, the brewing feud between Maria and Ed Piper finally erupted in bloodshed. Robert St. Marie, a recent Jones employee, was shot and killed by Ed Piper in the shadow of Weaver's Needle. As detailed in an earlier chapter, Piper was absolved of the killing (self defense) and set free. It was a senseless killing, like many others which had taken place in those mountains, and intensified the feelings between the warring camps. Fortunately, though, no more blood was spilled. Two years later, Piper died of natural causes, leaving Maria the heir to Weaver's Needle and the surrounding turf. But, she, too, was running out of time. In 1963, not long after Vance Bacon had fallen to his death while in the employ of Maria, she gradually reduced her prospecting activities, eventually abandoning the camp and drifting out of the Superstitions, swallowed up somewhere in the rites of conventional living, but never, I'm sure, abandoning her belief that the gold was right there where she thought it would be. It was still there in that hollowed-out room of rock, deep in the bowels of Weaver's Needle, high above the Superstition desert floor.

Bibliography

(The most extensive bibliography on the Lost Dutchman Mine and Superstition Mountain ever assembled—more than 500 entries!)

Booklets, Books, Manuscripts, Pamphlets

Aiton, Robert A., *The Lost Dutchman* (pamphlet), Lost Dutchman Mining Corp., Phoenix, Arizona, 1920.

Allen, Robert Joseph, *The Story of Superstition Mountain and the Lost Dutchman Gold Mine,* New York, Pocket Books, 1971.

Arizona Bureau of Mines, *Superstition Gold,* Phoenix.

Arizona Department of Economics and Development, *Ghost Towns and Lost Treasures in Arizona* (pamphlet), Phoenix.

Arizona Guide Series: *Arizona,* New York, Hastings House, 1940.

Arnold, Oren, *Sutter's Gold,* Phoenix, Arizona Printers, 1934.

Arnold, Oren, *Superstition's Gold,* Phoenix, Arizona Printers, 1934, 1946.

Arnold, Oren, "Ghost Gold," included in *Sun In Your Eyes,* University of New Mexico Press, Albuquerque, 1947.

Arnold, Oren, *Ghost Gold,* San Antonio, Texas, The Naylor Company, 1954.

Arnold, Oren, "Stay Away From Up There," included in *Hidden Treasures in the Wild West,* Abelard-Schuman Co., New York, 1966.

Arnold, Oren, *Mystery of Superstition Mountain,* Irvington-on-Hudson, New York, Harvey House, Inc., 1972.

Attebury, James D., *A Story of the Superstition Mountains: Fact and Legend,* Saint Clair County Historical Society, Osceola, Mo., 1954.

Ballantine, Bill, "Mountain Bewitched," included in *High West,* Rand McNally, Chicago.

Barnard, Barney, *The Story of Jacob Walzer and His Famous Hidden Mine (The Lost Dutchman),* Mesa, Arizona, Mesa Tribune, 1954.

Barnard, Barney, and Higham, Charles F., *Superstition Mountain and Its Famed Dutchman's Lost Mine,* Mesa, Arizona, 1952.

Barnard, Barney, and Higham, Charles F., *Superstition Mountain and Its Famed Dutchman's Lost Mine,* Apache Del Superstition, Apache Junction, Arizona, 1952.

Black, Harry G., *The Lost Dutchman Mine: A Short Story of a Tall Tale,* Boston, Branden Press, 1975.

Black, Jack, "Lost Dutchman Mine," included in *Western Treasures Notebook of Lost Mines and Buried Treasures,* Ames Publishing Co., Tarzana, Ca., 1966.

Black, Jack, "Peralta's Eight Mines," *Ibid.*

Blair, Robert, *Tales of the Superstitions: The Origins of the Lost Dutchman Legend,* Tempe, Arizona, Arizona Historical Foundation, 1975.

Botkin, B.A., "The Lost Dutchman Mine," included in *A Treasure of Western Folklore,* Crown Co., New York, 1951.

Burbridge, Jonathan S., *Arizona's Monument to Lost Mines,* n.p., Reno, Nevada and San Bernardino, Calif. (?), 1969.

Burns, Mike, *The Legend of Superstition Mountain,* ed. and rev. by Loyde E. Kingman, Phoenix, A Truman Helm, 1925.

Burns, Mike, (aka Hoomothya-Wet-Nose), *The Legend of Superstition Mountain,* new illustrated edition, Phoenix, Arrowhead Studios, 1935.

Clark, Howard D., "The Lost Dutchman Mine," included in *Lost Mines of the Old West*, Ghost Town Press, Buena Park, Calif., 1946.

Cleator, Philip Ellaby, "Gringo Gold," included in *Treasure for the Taking*, Robert Hale, London, 1960.

Colten, James, *Echoes of a Legend*, Apache Junction, Arizona, Apache Printing, 1977.

Colten, James, *The Old Apache Trail*, Apache Junction, Arizona, Apache Printing, 1980.

Conaster, Estee, *The Sterling Legend, The Facts Behind the Lost Dutchman Mine*, Dallas, Texas, Ram Publishing Company, 1972.

Conrotto, Eugene L., *Lost Desert Bonanzas*, Palm Desert, California, Desert-Southwest Publishers, 1963.

Cookridge, E. H., *The Baron of Arizona*, New York, John Day Company, 1967.

Coolidge, Dane, *Arizona Cowboys*, New York, E. P. Dutton and Co., 1938.

Corle, Edwin, *The Gila*, New York, Rinehart and Co., 1951. Repr. by University of Nebraska Press, Lincoln, 1967.

Dahlmann, John L., *A Tiny Bit of God's Creation*, (Sketches by Ted DeGrazia), Tempe, Arizona, Reliable Reproductions, 1979.

d'Autremont, Hugh, *West of Dawn*, New York, Exposition Press, 1971.

Davis, Gregory, *A Don in Washington, Interviews of Fred Guirey* (in manuscript form), March 1976.

DeGrazia, Ted, *DeGrazia and His Mountain, the Superstition*, Tucson, Arizona, University of Arizona Press, 1972.

Drago, Harry Sinclair, *Lost Bonanzas*, Dodd, Mead and Co., New York, 1966. Repr. by Bramhall House, New York, 1972.

Dunning, Charles H., and Peplow, Edward H., Jr., *Rocks to Riches*, Phoenix, Southwest Publishing Co., 1959.

Ely, Sims, *The Lost Dutchman Mine*, New York, William Morrow and Co., 1953.

Ferguson, Robert G., *Lost Treasures, The Search for Hidden Gold*, New York, Vantage Press, 1957.

Fish, Frank L., "The Lost Dutchman," included in *Buried Treasure and Lost Mines*, Amador Publishing Co., Chino, Calif., 1961, 1963, 1966, 1970.

Francis, Marilyn, *Thunder in the Superstitions*, Phoenix, Arizona, Firebird, 1959.

Gardner, Earl Stanley, *Hunting Lost Mines by Helicopter*, New York, William Morrow and Co., 1965.

Garman, Robert L., *Mystery Gold of the Superstitions*, Mesa, Arizona, Lane Printing and Publishing Co., 1975.

Gentry, Curt, *The Killer Mountains*, New York, New American Library, 1968.

Granger, Byrd Howell, *A Motif Index for Lost Mines and Treasures*, Tucson, Arizona, University of Arizona Press, 1977.

Hammond, Vern, "The Lost Dutchman Mine," included in *Famous Lost Mines of the Old West*, True Treasure Publications, Conroe, Texas, 1971.

Harding, Albert, *Dutchman's Gold*, Peoria, Arizona, Amric Books, 1982.

Helm, Thomas, "Dutchman's Gold," included in *Treasure Hunting around the World*, Dodd, Mead and Co., 1960.

Higham, Charles F., *The True Story of Jacob Walzer*, El Paso, Texas, McMath Company.

Higham, Charles F., *Superstition Mountain and Its Famed Dutchman's Lost Mine*, El Paso, Texas, McMath Printers, 1946.

Ives, Burl, *Tales of America*, Cleveland, Ohio, World Publishing Co., 1954.

Jennings, Gary, *The Treasure of the Superstition Mountains*, New York, W. W. Norton and Co., 1973.

Kitt, Edith Stratton, "Hunting a Mine," included in *Pioneering in Arizona,* Arizona Historical Society, Tucson, 1964.

Krippene, Ken, "Old Snowbeard's Lost Mine," included in *Buried Treasure,* New York, Perma Book Co., 1950.

Larson, Earnest W., *The Peralta Cache,* Stockton, Calif., Mavern Co., 1977.

Lee, Robert E., *The Making of the Motion Picture "The Lost Dutchman Mine,"* San Diego, D. Martin, 1974.

Lesure, Thomas B., "Trek for Lost Gold," included in *Adventures in Arizona,* Naylor Co., San Antonio, Texas.

Lewis, Jeffrey William, *Perception of Desert Wilderness by the Superstition Wilderness User,* Tucson, Arizona, University of Arizona Thesis, 1971.

Lively, William Irven, *The Apache Trail,* Phoenix, Arizona, Helm Publishers, 1945.

Lively, William Irven, *Fingers of Fire,* Phoenix, Arizona, 1948.

Lively, William Irven, *The Mystic Mountains; A History of the Superstition Mountains,* Phoenix, Arizona, 1955.

Lost Dutchman Mining Corporation, *The Lost Dutchman,* Phoenix, Arizona, 1920.

Lovelace, Leland, *Lost Mines and Hidden Treasure,* San Antonio, Texas, The Naylor Company, 1956.

Lovelace, Leland, *Lost Mines and Hidden Treasure,* New York, Ace Books, 1972.

Luca, Das, *Ghosts of Superstition Mountains,* unpublished manuscript, Apache Junction, Arizona, no date.

Marlowe, Travis, *Superstition Treasures,* Phoenix, Tyler Printing Co., 1965.

McCallister, R. W., *Lost Mines of California and the Southwest,* Los Angeles, Thomas Brothers, 1953.

McClintock, James H., *Arizona: Prehistoric-Aboriginal-Pioneer-Modern,* S. J. Clarke Publishing Co., Chicago, Vol. II, 1916.

Miller, Joseph, "Superstition Lore," included in *Arizona: The Last Frontier,* Hastings House, New York, 1956.

Mitchell, John D., *Lost Mines of the Great Southwest,* Phoenix, Press of the Journal Co., Inc., (Rose and Allison), 1933. Repr. by Rio Grande Press, Glorieta, N.M., 1970, 1971.

Mitchell, John D., *Lost Mines and Buried Treasure Along the Old Frontier,* Palm Desert, Calif., Desert Magazine Press, 1953. Repr. by Rio Grande Press, Glorieta, N.M., 1970, 1971.

Morrow, Albert Erland, *Famous Lost Gold Mines of Arizona's Superstition Mountains,* Kansas City, Kansas, n. p., 1957.

Mueller, Karl von, *Treasure Hunter's Manual,* Dallas, Texas, Ram Publishing Co., vol. 6., 1961, vol. 7, 1972.

Mueller, Karl von, *Encyclopedia of Buried Treasure,* Weeping Water, Neb., Exanimo Press, 1965.

Munch, Theodore, W., and Winthrop, Robert D., *Thunder on Forbidden Mountain,* Philadelphia, Pa., The Westminster Press, 1976.

Nathan, Robert, *The Mallot Diaries,* New York, Alfred A. Knopf, 1965.

Nelson, Dick and Sharon, *Hiker's Guide to the Superstition Mountains,* Glenwood, N.M., Tecolote Press, 1978.

Palmer, Ralph F., *Doctor on Horseback,* Mesa, Arizona, Mesa Historical Society, 1979.

Paris, Val., *Superstition Phantom,* Apache Junction, Arizona, unpublished manuscript, 1976.

Peck, Ann Merriam, *Southwest Roundup,* New York, Dodd, Mead and Co., 1950.

Penfield, Thomas, "Lost Mines," included in *Lost Treasure Trails,* New York, Grosset & Dunlap, 1954.

Penfield, Thomas, *Dig Here!* San Antonio, Texas, The Naylor Co., 1962, 1966, 1967, 1968, 1971.

Penfield, Thomas, *Buried Treasure and Where to Find It*, New York, Tempo Books, 1969.

Penfield, Thomas, Lost Dutchman and Peralta Mines articles included in *Directory of Buried or Sunken Treasures and Lost Mines of the United States*, Conroe, Texas, True Treasure Publications, 1971.

Penfield, Thomas, *A Guide to Treasure in Arizona*, Conroe, Texas, True Treasure Publications, 1973.

Pine, Harold, *How I Found the Lost Dutchman Mine*, manuscript, in files of Arizona Historical Society, Tucson, 1959.

Powell, Donald M., *The Peralta Grant*, Norman, Okla., University of Oklahoma Press, 1960.

Probert, Thomas, *Lost Mines and Buried Treasures of the West*, Berkeley and Los Angeles, Calif., University of California Press, 1977.

Rascoe, Jesse Ed, *Western Treasures, Lost and Found*, Toyahvale, Texas, Frontier Book Co., 1961.

Rascoe, Jesse Ed, *The Golden Crescent, the Southwest Treasure Belt*, Toyahvale, Texas, Frontier Book Co., 1962.

Rascoe, Jesse Ed, "The Lost Dutchman," included in *More Western Treasures*, Toyahvale, Texas, Frontier Book Co., 1962. Repr. in *Some Western Treasure Trails*, Cisco, Texas, Frontier Book Co., 1964.

Rascoe, Jesse Ed, "Superstition Silver," included in *Old Arizona Treasures*, Fort Davis, Texas, Frontier Book Co., 1970.

Robinson, Will H., *When the Red Gods Made Men*, Phoenix, Hubbard Printing Co., 1935.

Robinson, Will H., *Story of Arizona*, Phoenix, Berry Hill Co., n. d.

Rosecrans, Ludwig Groth, *Spanish Gold and the Lost Dutchman*, Mesa, Arizona, Lofgreen Printing and Office Supply Co., 1949.

Santschi, R. J., *Treasure Trails*, Oak Park, Ill., Century Press, 1938. Glen Ellyn, Ill., 1949.

Sayles, E. B. "Ted," *Fantasies of Gold*, Tucson, University of Arizona Press, 1968.

Schaefer, Lake Erie, Dutchman articles included in *Dead Men Do Tell Tales*, Chino, Calif., Amador Publishing Co., 1968.

Sheridan, Michael F., *Superstition Wilderness Guidebook: an Introduction to the Geology and Trails, Including a Roadlog of the Apache Trail and Trails from First Water and Don's Camp*, Tempe and Phoenix, Arizona, Sheridan, 1971, 1978.

Storm, Barry, *Trail of the Lost Dutchman*, Phoenix, Arizona, Goldwaters, 1939.

Storm, Barry, *Gold of the Superstitions*, Phoenix, Arizona, Sims Printing Company, 1940.

Storm, Barry, *Lost Arizona Gold*, Phoenix, Arizona, Storm-Mollet, 1953.

Storm, Barry, *Thunder God's Gold*, Phoenix, Arizona, Storm-Mollet, 1953.

Storm, Barry, *I Was Swindled by Red Movie Makers*, Phoenix, Storm-Mollet, 1954.

Storm, Barry, *Thunder God's Gold*, Chiriaco Summit, California, Stormjade, 1967.

Storm, Barry, *Mountains That Were God*, Indio, Calif., Stormjade Books, 1967.

Swanson, James and Kollenborn, Tom, *Superstition Mountain, A Ride Through Time*, Phoenix, Arrowhead Press, 1981.

Talbert, Ruth S., *Superstition Country and the Apache Trail*, Apache Junction, Arizona, n. p., 1972.

United States Department of the Interior, Bureau of Mines, *Mineral Investigation of the Superstition Wilderness and Contiguous RARE II Further Planning Areas, Gila, Maricopa, and Pinal Counties Arizona*, MLA 136-82 (open file report), Denver, Colorado, 1982.

United States Forest Service, *Wilderness Management Plan, Superstition Wilderness, Tonto National Park*, U.S. Department of the Interior, 1970.

United States Work Projects Administration, *Arizona, the Grand Canyon State: A State Guide*, New York, Hastings House, 1940.

United States Work Projects Administration, *Journal of the Pioneer and Walker Districts,* 1863-1865, New York, Hastings House, 1940.

Wagoner, Merry, *Treasure Tales Across the Counter,* Chicago, R&M Printing, 1965.

Ward, Don, "The Lost Dutchman Mine," included in *Bits of Silver,* New York, Hastings House, n. d.

Webber, Charles Wilkins, *The Gold Mines of the Gila,* at Huntington Library, San Marino, Calif., n. p., 1849.

Webber, Charles Wilkins, *Old Hicks the Guide,* at Huntington Library, San Marino, Calif., n. p., 1855.

Wilburn, John D., *The Riddle of the Lost Dutchman Mine,* Phoenix, Lofgreen Printing, 1975.

Wilburn, John D., *Superstition Gold Mines and the Lost Dutchman,* Mesa, Arizona, Lane Printing, 1978.

Winters, Wayne, "Arizona Pioneer Prospector Who Claimed to Have Discovered Lost Dutchman Mine," included in *Forgotten Mines and Treasures of the Great Southwest,* Tombstone Nugget Publishing Co., Tombstone, Ariz., 1972.

Wyllys, Rufus K., *Arizona: The History of a Frontier State,* Phoenix, Hobson & Herr, 1950.

Legal Documents
City, County or State Records

Thomas, Julia, Divorce action between Julia Thomas vs. Emil W. Thomas, no. 1367, Record Room of the Clerk of the Court, Maricopa County, Book M4, p. 519, Oct. 9, 1891.

Waltz, Jacob, declaration of intent to become a United States citizen, Circuit Court, Natchez, Mississippi, Nov. 12, 1848.

Waltz, Jacob, (listed as J. W. Walls), in the City and County of Sacramento, Calif. Census, Oct. 27, 1850, p. 171.

Waltz, Jacob, (listed as Jacob Walls), in the County of Los Angeles, Azusa Township Census, July 24, 1860, p. 278.

Waltz, Jacob, United States Citizenship application, Federal Court of the First District, Federal Court Records, Book no. 89, p. 155, Los Angeles, Calif., July 19, 1861.

Waltz, Jacob, filed on the "Gross Lode" claim, Pioneer District, Journal of the Pioneer and Walker Districts, p. 263. Also in Recorder's Office, Yavapai County Courthouse, Prescott, Arizona. September 21, 1863.

Waltz, Jacob, staked "Big Rebel" claim, Walnut Grove Mining District, Arizona, Yavapai County Recorder's Office, Book B, Prescott, Arizona, Sept. 14, 1864.

Waltz, Jacob, registered "Big Rebel" claim at *Ibid.,* Jan. 8, 1865.

Waltz, Jacob, signature on petition to Governor Goodwin for protection from Indians, State Library, Archives and Public Records, Phoenix, March 11, 1864.

Waltz, Jacob, in the Special Arizona Territorial Census of 1864, U.S. Senate Document no. 13, 89th Congress, 1865, Washington, D.C.

Waltz, Jacob, (listed as Jacob Waly), filed the "General Grant" claim in the Yavapai County Recorder's Office, Yavapai County Book of Mines, Book B, pp. 155-156, Prescott, Arizona, January 8, 1865.

Waltz, Jacob, "Claim for 160 Acres of Land," United States Land Office, Records, April 1868, Florence, Arizona.

Waltz, Jacob, filed "Improvement on 160 Acres of Land," Property Valuation and Assessment Rolls, Yavapai County, Arizona, April 1868, p. 53.

Waltz, Jacob, listed in United States Census for 1870 and 1880.

Waltz, Jacob, listed in Registration of Voters between 1871 and 1891, Maricopa County, Arizona.

Waltz, Jacob, filed affidavit, Arizona Recorder's Office, Book 1 of Misc. Records, Maricopa County, Phoenix, March 21, 1872, p. 107.

Waltz, Jacob, listed in the Great Register of Maricopa County, Phoenix, Arizona, 1876, 1882, 1886.

Waltz, Jacob, Deed to Andrew Starar, Arizona Records of Deeds, Maricopa County, Book 3, pp. 322-325, Phoenix, Arizona, Aug. 8, 1878.

Willing, George M. Jr., and Peralta, Miguel (agreement between), Arizona Recorder's Office, Yavapai County, Book I of Misc. Records, pp. 93-94, Prescott, March 12, 1874.

Maps

Arizona Bureau of Mines, *Geologic Map of Pinal County, Arizona*, Tucson, Arizona, Arizona Bureau of Mines, University of Arizona, 1959.

Arizona Bureau of Mines, *Geologic Map of Maricopa County, Arizona*, Tucson, Arizona, Arizona Bureau of Mines, University of Arizona, 1959.

Fenninger, A., *Old Timer's Map of the Superstition Mountains*, Wickenburg, Arizona, A. Fenninger, 1964.

Nelson, Richard and Sharon, *The Superstition Wilderness, Western Half*, Glenwood, N.M., Tecolote Press, Inc., 1980.

Stirrat, J. A., *Map of the Lost Dutchman and Other Legendary Mine Areas in the Superstition Mountains of Arizona*, El Paso, Texas, J. A. Stirrat, 1948.

U.S. Department of Agriculture, Forest Service, *Tonto National Forest Recreation Map (Arizona)*, Albuquerque, N.M., 1962.

U.S. Department of Agriculture, Forest Service, Region 3, *Mesa Ranger District, Tonto National Forest*, Mesa, Arizona, no date.

U.S. Department of Agriculture, Forest Service, *Superstition Wilderness*, Albuquerque, N.M., 1981.

U.S. Geological Survey, *Iron Mountain Quadrangle, Arizona (Topographic)*, Denver, Colorado or Washington, D.C., U.S. Geological Survey, 1948.

U.S. Geological Survey, *Florence Quadrangle, Arizona (Topographic)*, Denver, Colorado or Washington, D.C., U.S. Geological Survey, 1955.

U.S. Geological Survey, *Goldfield Quadrangle, Arizona (Topographic)*, Denver, Colorado or Washington, D.C., U.S. Geological Survey, 1956.

U.S. Geological Survey, *Roosevelt Quadrangle, Arizona (Topographic)*, Denver, Colorado or Washington, D.C., U.S. Geological Survey, 1964.

U.S. Geological Survey, *Weaver's Needle Quadrangle, Arizona (Topographic)*, Denver, Colorado or Washington, D.C., U.S. Geological Survey, 1966.

U.S. Bureau of Mines, Map of Superstition Wilderness included in report: *Mineral Investigation of the Superstition Wilderness and Contiguous RARE II Further Planning Areas, Gila, Maricopa, and Pinal Counties, Arizona, MLA 136-82*, Denver, Colorado, U.S. Department of the Interior, 1982.

Newspapers

Apache Sentinel, Apache Junction and Mesa, Arizona, June 12, 19, July 17, October 30, November 13, 20, 27, 1959. Jan. 1, April 8, 1960. March 27, April 10, 1964. Jan. 27, 1982.

Arizona Daily Citizen, Tucson, Arizona, Feb. 20, 1896.

Arizona Daily Gazette, Phoenix, Arizona, June 18, 1884. Oct. 27, 1891. Oct. 1893, Dec. 1893. Nov. 5, 1894. Aug. 20, 1920.

Arizona Daily Star, Tucson, Arizona, Feb. 25, 1912. Nov. 9, 1924. March 24, 1927. April 26, May 21, June 26, July 19, July 22, Dec. 21, 1931. Jan. 8, 9, 1932. Jan. 14, 1934. Jan. 31, 1935. April 10, 1936. Dec. 13, 1940. Jan. 25, June 15, 1946. Jan. 24, 1947. March 24, 1951. Jan. 18, 1956. Oct. 3, 1962. March 23, 1963. Feb. 18, March 14, 1966. Dec. 19, 1970. Nov. 27, 1971. April 16, 1975. July 8, 1981. Jan. 16, 1982.

Arizona Daily Sun, Flagstaff, Arizona, Jan. 18, 1931.

Arizona Daily Wildcat, Tucson, Arizona, April 16, 1975.

Arizona Enterprise, Phoenix, Arizona, June 9, Sept. 1, 1892.

Arizona Gazette, Phoenix, Arizona, June 19, 1884. Sept. 1, 1892. Aug. 22, 1895. Feb. 4, 28, 1896. Aug. 20, 1920. July 24, 1931.

Arizona Miner, Prescott, Arizona, Sept. 7-21, 1864.

Arizonan, Chandler, Arizona, July 26, 1940.

The Arizona Republic, Phoenix, Arizona, July 26, Dec. 13, 14, 15, 16, 17, 18, 19, 20, 21, 22, 23, 31, 1931. Jan. 4, 9, 10, April 16, Nov. 16, 1932. June 8, 1934. Jan. 8, 9, Feb. 10, April 10, 1935. Dec. 12, 1937. March 26, 27, 28, 1940. March 3, Aug. 17, Sept. 12, 1941. Jan. 26, April 6, June 15, 1947. Feb. 21, 22, 1948. April 15, 1951. Dec. 20, 22, 1953. June 24, 1959. May 16, 1965. July 11, 1967. Dec. 13, 1970. Dec. 6, 27, 1981. Jan. 15, 1982. Feb. 6, 1983.

The Arizona Republican, Phoenix, Arizona, Feb. 19, 1891.

Arizona Sentinel, Phoenix, Arizona, May 16, 1896.

Arizona Silver Belt, Miami, Arizona, Oct. 24, 1924. July 16, 1930.

Constitution-Morning Press, Lawton, Oklahoma, June 21, 1950.

Daily Oklahoman, Oklahoma City, Oklahoma, Jan. 14, 15, 16, 17, 18, 19, 1970.

Denver Post, Denver, Colorado, Nov. 13, 1936. July 31, 1949.

Dispatch, Casa Grande, Arizona, Nov. 30, 1934. March 8, 1971.

Evening Outlook, Santa Monica, Calif., Aug. 24, 1971.

Los Angeles Herald-Examiner, Los Angeles, Calif., Sept. 10, 1933. April 13, 1941. Feb. 27, 1972.

Los Angeles Times, Los Angeles, Calif., March 6, 1932. Feb. 4, 1934. April 10, 1950. May 31, 1970.

Mesa Journal-Tribune, Mesa, Arizona, Jan. 12, 1931.

Oakland Tribune, Oakland, Calif., April 17, 1949.

The Observer, St. John's Arizona, Aug. 29, 1913.

The Phoenix Gazette, Phoenix, Arizona, 1886. 1893. Jan. 8, 9, July 26, 1935. May 7, 1941. March 22, 1942. April 29, 1966. Dec. 25, 1970. Jan. 15, 1972. Jan. 6, 1975. Dec. 15, 1977. Jan. 12, 25, Feb. 15, 1980.

Phoenix Herald, Phoenix, Arizona, Dec. 5, 1879. 1884. Feb. 20, 26, Oct. 25, 1891.

Prescott Evening Courier, Prescott, Arizona, July 10, 1931. June 29, 1946.

The Prospect, Prescott, Arizona, Sept. 28, 1901.

Rocky Mountain News, Denver, Colorado, Jan. 11, 1914. Jan. 24, 1947.

San Diego Union, San Diego, Calif., Dec. 22, 1919.

Saturday Review, Phoenix, Arizona, Nov. 17, 1894. Aug. 24, 1895.

Scottsdale Progress, Scottsdale, Arizona, March 10, 1982.

Tombstone Prospector, Tombstone, Arizona, Dec. 30, 1900.

Tucson Daily Citizen, Tucson, Arizona, Feb. 20, 1896. March 24, 1921. Feb. 14, 1924. March 24, 1927. June 25, 1931. Jan. 8, 1932. June 8, Sept. 29, 1934. Oct. 10, 1961. Nov. 26, 1971.

The Wall Street Journal, New York, N.Y., Oct. 18, 1971.

The Weekly Arizona Miner, Prescott, Arizona, Feb. 15, 1878.

Periodicals

Ackerman, R. O., "Madman of the Superstitions," *True West,* Jan.-Feb. 1956, pp. 22-24.

Ackerman, R. O., "Madman of the Superstitions," *Gold,* vol. 1, no. 1, 1969, pp. 62-64.

Alan, G. Arthur, "Death Guards a Lost Mine," *Western Yarns,* March 1943, pp. 84-88.

Allen, Robert Joseph, "The 100-year Mystery of the Murdering Mountain," *True,* Feb. 1962, pp. 97-101.

Allen, Robert Joseph, "Hidden Ledge of Gold," *Point West*, Feb. 1963, pp. 5, 33-34.

Andreeva, Tamara, "Here Lies Buried Treasure," *True West*, Summer 1953, pp. 16, 54.

Andreeva, Tamara, "Here Lies Buried Treasure," *Old West*, Winter 1964, pp. 30-31.

Andreeva, Tamara, "Here Lies Buried Treasure," *Gold*, vol. 1, no. 1, 1969, pp. 42-43.

"Arizona Landmark," *Arizona Days and Ways*, June 5, 1960, p. 11.

"Arizona, Search for Lost Dutchman's," *Time*, June 22, 1959, pp. 17-18.

Arnold, Oren, "The Lost Dutchman," *Wide World*, March 1933, April 1933.

Arnold, Oren, "Gold in the Mountains," *Arizona Highways*, Feb. 1938, p. 12, 19, 26.

Arnold, Oren, "I Hunt Ghost Gold," *Saturday Evening Post*, Dec. 28, 1946, pp. 9-11.

Arnold, Oren, "Lure of the Superstition Mountains: Lost Dutchman Mine," *Coronet*, April 1956, pp. 52-55.

Austin, Mary, "Treasure in the Southwest," *Frontier Times*, Feb. 1944, p. 222.

Bagwell, Mary L., "An Old Timer Sheds New Light on a Famous Lost Lode: The Lost Dutchman Mine," *Desert*, Jan. 1954, pp. 18-19.

Bailey, Tom, "Looking for a Lost Mine?" *True West*, vol. 3.

Balliett, W., "The Lost Dutchman Mine," A Review of the book by Sims Ely, *Saturday Review*, Dec. 12, 1953, p. 27.

Bannock, Charles, "Gold Fever," *Peoples Magazine of Arizona*, March 1939, p. 6, 25.

Barnard, Barney, "The Truth About the Dutchman's Lost Mine," *True West*, July-Aug. 1955, pp. 16-18, 36-40.

Bassler, M., "The Ghost Bell," *Progressive Arizona*, Dec. 1928, p. 16.

Beck, L. J., "Mystery of the Dutchman's Lost Mine," *True West*, Nov.-Dec. 1955, pp. 38-39.

Beck, L. J., "Mystery of the Dutchman's Lost Mine," *Gold*, vol. 1, no. 1, 1969, p. 81.

Bell, L. J., "Dutchman's Gold Not Really a Mine?" *True West*, May-June 1961, pp. 46-47.

Bernhard, Morris A., "Know Your Treasure Hunting," *The West*, Sept. 1966, p. 46.

Bernhard, Morris A., "Pieces of Eight," *Gold*, vol. 3, no. 2, 1971, p. 67.

Black, William J., "Mystery of the Dutchman's Lost Mine," *True West*, Nov.-Dec. 1955, p. 27.

Black, William J., "Mystery of the Dutchman's Lost Mine," *Gold*, vol. 1, no. 1, Annual, 1969, p. 79.

Blair, Robert, "The Lost Dutchman Legend," *Desert*, April 1966, pp. 14-16.

Blair, Robert, "Beware of Swindlers," *Desert*, Aug.-Sept. 1966, p. 63.

Blair, W., "Ride to Angel Springs," *Outdoor Arizona*, Oct. 1971.

Brown, Art, "Still More Lost Than Found," *Desert*, July 1966, p. 38.

Brown, Dr. Orville Harry, "Skeleton Cave," *Arizona, Old and New*, July 1928, pp. 22-24, Aug. 1928, pp. 16-18, Sept. 1928, pp. 31-32.

Burbridge, Jonathan S., "The Lost Walker Map," *Gold*, Summer 1971, p. 27.

Burbridge, Jonathan S., "The Superstitions: Mountains of Foam," *Treasure Hunter*, vol. 7, no. 3, pp. 10-13.

Burnham, Ben R., "Mystery of the Dutchman's Lost Mine," *True West*, Nov.-Dec. 1955, p. 27.

Burnham, Ben R., "Mystery of the Dutchman's Lost Mine," *Gold*, vol. 1, no. 1, 1969, p. 79.

Burns, Mike, "The Legend of Superstition Mountain," *Progressive Arizona*, March 1926, p. 5

Carson, K., "Why Can't Anyone Find Dr. Thorne's Lost Gold Mine?" *The West*, Sept. 1974.

Castillo, Juan, "The Hermit of the Superstitions," *Magazine Tucson*, Oct. 8, 1952, pp. 30-31.

Chambliss, Catherine, "The Lure of the Superstitions," *Arizona Highways*, Nov. 1944, pp. 31-34.

Charles, P., "Lure of the Superstitions," *Arizona Wildlife Sportsman*, Jan. 1953, pp. 30-33.

Chesney, W. D., "Pieces of Eight," *Gold*, Summer 1971, p. 44.

Clark, Charles M., "History of the Lost Dutchman Gold Mine," *Arizona Mining Journal,* March 30, 1925, pp. 11-12.

Collier, Bill W., "Mystery of the Dutchman's Lost Mine," *True West,* Nov.-Dec. 1955, pp. 37-38.

Conniston, Ralph, "Sixty Billion Dollars in Lost Gold," *True West,* June 1955, pp. 6-9, 26.

Corbin, Helen M., "Piper of the Superstitions," *True West,* July-Aug. 1965, pp. 36-37.

Cox, Quinton T., "A Challenge," *Treasure World,* Oct.-Nov. 1971, p. 10.

Cox, Quinton T., "Pieces of Eight," *Gold,* vol. 3, no. 2, 1971, pp. 51, 66.

Cox, Quinton T., "More on the Lost Dutchman," *Treasure Hunter,* vol. 3, no. 6, p. 6-7, 21.

Daley, Bob, "Lost Dutchman," *Treasure World,* June-July 1973, p. 13.

David, Lester, and Lewis, Irwin, "Found Key to World's Richest Gold Mine." *Mechanix Illustrated,* Feb. 1942, pp. 34-35, 171.

Davis, Gregory E., "More on the Truth About Jacob Waltz," *Treasure Hunter,* Sept.-Oct. 1970, p. 17.

Davis, Gregory E., "Superstition Mountain Journal," Superstition Mountain Historical Society, Inc., vol. 1, 1981.

"Death Guards Lost Gold of Dutchman Mine," *Popular Mechanics,* July 1938, pp. 24-25.

Denton, Jon, "The Gold Diggers," *Oklahoma Orbit,* Jan. 14, 1970 (supplement to the *Sunday Oklahoman*).

"Desert Place Names," *Desert,* Feb. 1938, p. 23.

DeWald, B., "Trek Into the Fabled Superstition Mountains," *Arizona Days and Ways,* Jan. 5, 1958, pp. 16-19.

DeWald, B., "Threading the Needle," *Arizona Days and Ways,* Jan. 8, 1961, pp. 27-31.

Dillon R., "Fish Creek—Oasis on the Apache Trail," *Phoenix,* Jan. 1974.

Dillon, R., "Superstition Trail to Hieroglyphic Spring," *Outdoor Arizona,* Oct. 1975.

Duncan, Virginia, "Trekking for Treasure," *Desert,* April 1938, pp. 4-6.

Earnest, G. A., "Mystery of the Dutchman's Lost Mine," *True West,* Nov.-Dec. 1955, p. 28.

Earnest, G. A., "Mystery of the Dutchman's Lost Mine," *Gold,* vol. 1, no. 1, 1969, p. 80.

Edmiston, Ray L., "La Sombrera and the Mountain of Gold," *Prospector-Outdoorsman,* July 1960, p. 2, 3, 5.

Emerson, Lee, "Petroglyphs of Ancient Man," *The Indian Historian,* Spring 1971, pp. 4-8.

Espinoza, Dan, "The Superstitions: Arena of Death," *The Sheriff,* Aug. 1960, pp. 31-35.

Farnbach, F., "Hiking the Dutchman's Trail," *Outdoor Arizona,* March 1971.

Garman, Robert L., "Quest of the Peralta Gold," *Desert,* Feb. 1953, pp. 22-24.

Goldberger, Herb, "The Lost Dutchman Found?" *Treasure Adventure,* vol. 1, no. 4, 1962, pp. 58-62.

Goodman, V. H., "The Lost Dutchman Mine," *Desert,* June 1954, p. 27.

Grey, Don, "Mystery of the Dutchman's Lost Mine," *Gold,* vol. 1, no. 1, 1969, p. 82.

Guirey, Fred, "Hospitality Western Style—The Story of the DONS CLUB of Phoenix, Arizona," *Arizona Highways,* March 1967.

Gustell, Chris, K., "Mystery of the Dutchman's Lost Mine," *Gold,* vol. 1, no. 1, 1969, p. 82.

Hammond, Vern, "The Lost Dutchman - Today," *True Treasure,* March-April 1968, pp. 9-17.

Hammond, Vern, "The Lost Dutchman Mine," *Western Fiction,* Dec. 1970-Jan, 1971, pp. 42-47.

Hawkins, Amos, (as told to Jean Long), "I Packed Al Morrow Out of the Superstitions," *True West,* Dec. 1972, pp. 16-18, 42-43.

Herlick, Martin H., "Charles Williams and the Superstition Mountain Gold Mystery," *Peoples Magazine of Arizona,* Nov. 1939, pp. 3-5.

Higham, Charles F., "The Peralta Coat of Arms," *Phoenix Spotlight,* Feb. 1-8, 1952.

Higham, Charles F., "Lost Gold Mine Maps Deal," *Ibid.,* June 5-12, 1953.

Higham, Charles F., "The Lost Dutchman Gold Mine," *Ibid.*, Jan. 1-8, 1954.
Higham, Charles F., "The Lost German Mine," *Ibid.*, May 1954.
Higham, Charles F., "Jacob Walzer's Grave," *Ibid.*, June 1954.
Higham, Charles, F., "Geronimo's Gold Hoards," *Ibid.*, July 1954.
Higham, Charles F., "The Mystery of Roger's Canyon," *Ibid.*, Aug. 1954.
Higham, Charles F., "Jacob Waltzer's Discovery," *Ibid.*, Nov. 19-26, 1954.
Higham, Charles F., "The True Story of Jacob Walser and His Famous Hidden Gold Mine," *Arizona Sheriff,* April 1973.
Hill, B., "Last Waltz in Phoenix," *Arizona,* Sept. 30, 1979.
Hilton, J. W., "Crystals in the Shadow of the Superstitions," *Desert,* March 1941, pp. 15-18.
Howe, Carl, "Did the Dutchman Find Montezuma's Treasure?" *True West,* Jan.-Feb. 1957, pp. 18-19, 41.
Howe, Carl, "Did the Dutchman Find Montezuma's Treasure?" *Gold,* vol. 1, no. 1, 1969, pp. 40-41, 61.
Hurley, Gerald T., "Buried Treasure Tales in America," *Western Folklore,* vol. 10, no. 3, 1951.
James, G. W., "A Legend of the Pimas: the Indian Story of Superstition Mountain," *Arizona,* Feb.-March 1917, p. 11.
Johnson, B., "The Four-Colored, Four-Directional Lighting From Within, An Apache Medicine Woman's Tale of the Superstition Mountains," *Arizona,* March 3, 1968.
Johnson, B., "The Snowbird's Guide to Snowbeard's Gold," *Arizona,* March 15, 1970.
Jones, Jerry, "Mystery of the Dutchman's Lost Mine," *True West,* Nov.-Dec. 1955, pp. 27-28.
Jones, Jerry, "Mystery of the Dutchman's Lost Mine," *Gold,* vol. 1, no. 1, 1969, pp. 79-80.
Judd, Ira B., "Legends of Arizona," *Point West Magazine of Arizona,* Oct. 1961.
Kalderberg, Phil, "Mystery of the Dutchman's Lost Mine," *Gold,* vol. 1, no. 1, 1969, pp. 82-83.
Kildare, Maurice, "The Dutchman's Three Lost Mines," *True Treasure,* July-Aug. 1973, pp. 34-36, 38-39.
Leatham, N., "A Nearby Hike to a Primitive World," *Arizona Days and Ways,* May 13, 1962, pp. 5-7.
Leatham, N., "Treasure the Dutchman Couldn't Hide," *Arizona Days and Ways,* March 21, 1965, pp. 40-43.
Leatham, N., "Getting High in the Superstitions," *Arizona,* Feb. 1, 1976.
"Legend of a Name," *The Oasis,* August 10, 1893.
"The Legend of Superstition Mountain," *Magazine Tucson,* Sept. 1950.
Lesure, Thomas, B., "Trek for Lost Gold," *Desert,* March 1954, pp. 19-20.
Lesure, Thomas B., "Pathway to Lost Mine Trek," *Desert,* March 1960, p. 33.
Levene, Gene, "Treasure Notes," *Treasure Hunter,* vol. 2, no. 1, p. 8.
Lewis, A. L., "Mystery of the Dutchman's Lost Mine," *True West,* Nov.-Dec. 1955, p. 28.
Lewis, A. L., "Mystery of the Dutchman's Lost Mine," *Gold,* vol. 1, no. 1, 1969, p. 80.
Lewis, Jack, "The Lost Dutchman - 1961," *Treasure Adventure,* Summer 1961, pp. 8-10, 50-52.
Long, Jean, "I Packed Al Morrow Out of the Superstitions," (Amos Hawkins' story), *True West,* Nov.-Dec. 1972, pp. 16-18, 42-43.
"Lost Dutchman Mine," *Newsweek,* June 18, 1962, p. 29.
"Lost Dutchman Reported Found: Twice in Different Places," *Western Treasures,* Aug. 1966, p. 8.
"Lost Quartz Vein of the Tonto Apache Indians," *Desert,* Feb. 1942.
"Lure of Gold in Fabled Superstition Mountains of Arizona Begets Murder: Lost Dutchman Claims Another Victim," *Western Mining and Industrial News,* July 1959, p. 1.

Marranzino, Pasquale, "Gold-Smitten Rush to Aid Grandma in Hunt for Mine." *Western Folklore Quarterly*, April 1947, pp. 185-186.

Martin, Scudder, "Superstition Mine," *Overland Monthly*, April 1923, pp. 16-18, 40.

Maxor, E. J., "Killer Mine," *Western Tales*, April 1960, pp. 28-31, 58.

May, Tom, "Life on the Desert," *Desert*, Feb. 1955, pp. 22-23.

McClintock, James H., "Fighting Apaches," *Sunset*, Feb. 1907, pp. 340-341, 358-359.

McGee, Bernice, "The Other World of the Superstitions," *Old West*, Winter 1964, pp. 2-15, 54-60.

McGee, Bernice and Jack, "Invitation to a Ghost Walk," *True West*, March-April 1966, pp. 6-17, 49-50.

McGee, Bernice and Jack, "Are the Peralta Stone Maps a Hoax?" *Frontier Times*, May 1973, pp. 6-13, 47-49, 52-57.

Millen, Charles, "Treasure Hunters, Attention," *Gold*, Summer 1971, p. 42.

Miller, Dean C., "Treasure Notes," *Treasure Hunter*, vol. 2, no. 1, p. 8.

Miller, Ralph, "Mystery of the Dutchman's Lost Mine," *True West*, Nov.-Dec. 1955, p. 38.

Miller, Ralph, "Mystery of the Dutchman's Lost Mine," *Gold*, vol. 1, no. 1, 1969, pp. 80-81.

Mitchell, John D., "The Lost Mexican Mine," *Arizona Mining Journal*, vol. 18, no. 24, 1935, p. 5

Mitchell, John D., "Lost Apache Gold Mine," *Desert*, April 1940, p. 18.

Mitchell, John D., "Lost Dutchman Mine," *Desert*, March 1941, pp. 27-28.

Mitchell, John D., "The Lost Dutchman Gold Mine," *Mining World*, vol. 12, no. 4, 1950, p. 18, 66.

Mitchell, John D., "Lost Apache Gold Mine," *Desert*, Jan. 1967.

Monagan, George, "Dutchman's Lost Mine Found?" *True West*, March-April 1956, pp. 26-28.

Monagan, George, "Dutchman's Lost Mine Found?" *Gold*, vol. 1, no. 1, 1969, pp. 16-17, 61.

Montagu, Ashley, "Ales Hrdlicka, 1869-1943," *The American Anthropologist*, 1944.

Moore, H. R., "Superstition Picnic," *Arizona Days and Ways*, July 16, 1961, pp. 24-27.

Morando, B., "The Murderous Lost Dutchman Mine," *Lost Gold: Hidden Treasure of the West*, Summer, 1964, pp. 54-58.

Oliver, Harry, "Arizona Lost Gold," *Desert Rat Scrapbook*, packet 4, pouch 1, p. 4.

Oliver, Harry, "Thar's Gold in Them Thar Hills," *Ibid.*, packet 1, pouch 4.

Oliver, Harry, "Texans Report Finding Lost Dutchman Mine," *Ibid.*, packet 1, pouch 11.

O'Neill, William Buckey, Article on the Lost Dutchman Mine, *Argonaut Magazine*, circa 1898.

Ortegas, Manual, "Treasure Notes," *Treasure Hunter*, vol. 2, no. 1, p. 8.

Peplow, Edward H. Jr., "The Legend of Superstition Mountain, Arizona's Magic Mountain," *Outdoor Arizona*, Oct. 1973, pp. 25-28, 37-40.

"Phoenix Finds," *Phoenix*, April 1982.

Pitts, John B., "Mystery of the Dutchman's Lost Mine," *True West*, Nov.-Dec. 1955, pp. 28, 37.

Pitts, John B., "Mystery of the Dutchman's Lost Mine," *Gold*, vol. 1, no. 1, 1969, p. 80.

Pounds, J. W., "Lost Dutchman Mine," *Treasure Hunter*, vol. 2, no. 3, p. 3.

Pounds, J. W., "Lost Dutchman Mine Found?" *Ibid.*, p. 10, 16.

Pounds, J. W., "The Dutchman Found!" *Ibid.*, vol. 5, no. 3, p. 6, 10.

Pounds, J. W., "Does Death Guard Superstition Gold?" *Ibid.*, pp. 4-5, 21.

Reed, Allen C., "Trek for Gold," *Arizona Highways*, March 1951, pp. 2-3, 26-27.

Reid, Peter, "Is the Treasure Really There?" *Gold*, Summer 1971, p. 42, 44.

"The Richest Mine in the World," *Arizona Sheriff*, Nov.-Dec. 1976.

Ricketts, C. E., "Pieces of Eight," *Gold,* Summer 1971, p. 42, 44.

Rieseberg, Harry E., "Lost Mines of Arizona," *True Frontier,* Sept. 1970, p. 17.

Roberts, John, "The Deadly Gold of the Superstitions," *Fury,* Sept. 1956, pp. 36-37, 66-69.

Rogers, Don, "Stay Away from Up There," *Family Circle,* Feb. 18, 1938, pp. 10-11, 16-17, 19.

Rose, Milton F., Letter published in *Western Treasures,* April 1972, p. 6.

Ross, Irwin, "The Golden Mine That Dares to be Found," *Elks Magazine,* April 1973, pp. 14-16.

Roush, Roy W., "Old Peralta Map Found?" *Treasure Hunter,* vol. 4, no. 1, p. 11.

"Sale of 'Authentic' Lost Dutchman Maps," *Western Folklore Quarterly,* Oct. 1950, pp. 78-79.

Senzee, W., "The Pied Piper of Crooked Mountain," *Outdoor Arizona,* May 1976.

Shaw, Victor, "The Lost Dutchman - Fact or Fable?" *Earth Science Digest,* vol. 1, 1947, pp. 3-9, 23, 25-26, 32, 37, 39-40.

Shaw, Victor, "Superstition Mountain Geology," *Mineralogist,* vol. 15, no. 10, 1947, pp. 507-510. Reprint in *Ibid.,* vol. 24, no. 1, 1956, pp. 8-10, 12.

Sheridan, Michael F., "Volcanic Geology Along the Western Part of the Apache Trail, Arizona," *Arizona Geological Society Southern Arizona Guidebook III,* 1968, pp. 227-229.

Small, Joe, "Dutchman Mine Lost Forever?" *Gold,* Spring 1972, p. 1, 5.

Sonneman, D. E., "What Is It?" *Treasure Adventure,* vol. 1, no. 4, 1962, p. 4.

Spanish, Johnny, "The Cave of Gold," *True Frontier,* May 1971, pp. 30-31, 40-41, 64.

Spanish, Johnny, "The Cave of Gold," *Ibid.,* Winter 1973-74, pp. 26-27, 46-49.

Starry, R. M., "One More Mystery for the Superstitions," *True West,* Nov.-Dec. 1974.

Stevens, Red, "Mystery of the Dutchman's Lost Mine," *True West,* Nov.-Dec. 1955, p. 27.

Stevens, Red, Reprint of above in *Gold,* vol. 1, no. 1, 1969, p. 29.

Stewart, A. J., "The $500,000,000 Treasure That's Up for Grabs," *Western Action,* Sept. 1960.

Stolley, Richard B., "Mysterious Maps to Lost Gold Mines: Superstition Mountains, Arizona," *Life,* June 12, 1964, pp. 90-94, 96.

Storm, Barry, "Soldier's Lost Vein of Gold," *Desert,* Jan. 1945, pp. 23-24.

Storm, Barry, "Lost Mines of the Peraltas," *Desert,* March 1945, pp. 25-28.

Storm, Barry, "Curse of the Thunder Gods," *Desert,* April 1945, pp. 10-12.

Storm, Barry, "Bonanza of the Lost Dutchman," *Desert,* May 1945, pp. 16-18.

Storm, Barry, "Rumors of Gold," *Desert,* June 1945, pp. 20-22.

Storm, Barry, "Mountain Treasure," *Desert,* July 1945, pp. 25-27.

Storm, Barry, "Storm Deciphers Peralta Maps," *Treasure Hunter,* vol. 3, no. 1, pp. 4-8, 16-22, 24.

Stuckless, J. S. and Sheridan, M. F., "Tertiary Volcanic Sequence in the Goldfield and Superstition Mountains, Arizona," *Geological Society of America Bulletin,* 1971.

Sykes, E. V., "Mystery of the Dutchman's Lost Mine," *True West,* Nov.-Dec. 1955, pp. 39-40.

Sykes, E. V., Reprint of above article in *Gold,* vol. 1, no. 1, 1969, p. 81.

"Take a Trek to the Superstition Wilderness Area," *Outdoor Arizona,* April 1977.

Taylor, Frank, "In the Shadow of the Needle," *Western Treasures,* March 1970, pp. 34-35.

Taylor, Frank, Reprint of above article in *The Best of Western Treasures,* annual, 1972, pp. 23-24.

Taylor, John A., "How the Dutchman Got His Gold," *True West,* Dec. 1956, p. 21, 36-37.

Taylor, John A., Reprint of above article in *Gold,* vol. 1, no. 1, 1969, p. 33, 92-93.

Taylor, R., "Lost Apache Gold," *Desert,* Oct. 1969.

Thayer, Victor I., "Pieces of Eight," *Gold,* vol. 3, no. 2, 1971, p. 4.

Thomas, Dale, "Treasure Enough for Everyone," *Western Treasures,* Summer 1963, pp. 22-25, 60, 62-65.

Thompson, W. C., "Pieces of Eight," *Gold,* vol. 3, no. 2, 1971, p. 67.

Thornton, Eric, "Curse of the Lost Dutchman," *Man's Cavalcade,* July 1957, pp. 36-39.

Thoroman, E. C., "Lost Gold of the Four Peaks," *Desert,* Nov. 1957, pp. 21-22.

Walker, C. Lester, "Where to Find Buried Treasure," *Harper's,* June 1947, p. 547.

Watson, Lou, "Legend of the Land: The Real Treasure of the Superstitions," *Arizona Wildlife Sportsman,* Nov. 1962, pp. 19-21.

Webster, I., "The Spectacular Superstitions," *Arizona Highways,* Nov. 1970.

Welch, James E., "Has the Lost Dutchman Been Found?" *True West,* March-April 1963, pp. 14-15, 59.

Whitaker, R. B., "Game Rangers on the Superstition Trail," *Arizona Days and Ways,* May 20, 1962, pp. 30-35.

Willson, Roscoe G., "Jose Varela Sees Dwarfs in the Superstition Mountains," *Arizona Days and Ways,* Dec. 31, 1961, pp. 30-31.

Willson, Roscoe G., "Superstition Subject of Another Book," *Arizona Days and Ways,* May 23, 1965.

Willson, Roscoe G., "Lost Dutchman Still Lost," *Arizona,* July 28, 1968.

Willson, Roscoe G., "The Little Peoples of the Superstitions," *Arizona,* Oct. 21, 1973, pp. 64-65.

Winters, Wayne, "Lost Dutchman Still Kills," *Prospector-Outdoorsman,* March 1960, pp. 4-5.

Index

A

Adams, Eva, 122
Adams, Jeff, 86
Agglomerate, 73, 74
Aiton, R. A., 95, 99, 100
Alsap, John T., 44
Andesite, 73, 74
Apache Highway, 75
Apache Indians, 20, 47-52, 62-65, 69, 84, 102
Apache Junction, 16, 17, 76, 91, 102
Apache Lake, 16
Apache Sentinel, 101-102
Apache Trail, 16, 17, 89
Arizona, 10, 11, 15, 19, 20, 36-38, 47, 55
Arizona Bureau of Mines, 73
Arizona Citizen, 25
Arizona Daily Star, 83, 89
Arizona Dept. Mineral Resources, 78, 80
Arizona Enterprise, 55
Arizona Gazette, 27, 30, 99, 100
Arizona Historical Society, 101
Arizona Journal-Miner, 69
Arizona Miner, 25
Arizona Republic, 84. 86, 88, 96, 102-103, 106
Arizona State Museum, 110
Arizona State University, 73
Arizona Territory, 25
Arizona Weekly Enterprise, 69
Associated Press, 104
Athabascan, 47
Aztec Indians, 9, 21, 52

B

Bacon, Vance, 46, 92, 127
Bailey, Audrey, 92
Bark, Jim, 61
Barkley, W. A., "Tex," 86
Barnard, Barney, 101
Basalt, 15, 74, 87
Battle of Skull Cave, 49
Better Half Cartoon, 105
Bicknell, P. C., 60-61
Big Rebel Claim, 40
Bilbry, Michael, 106
Biotite, 15
Bisbee, 74
Black Mesa, 74
Blacktop Mountain, 74, 118
Bley, Ron, 93
Blue Point, 98
Bluff Springs Mountain, 107, 108, 118
Bohen, Charles, 92
Boher, Grace, 88
Boies, Cal, 84
Boulder Canyon, 23, 66-67, 74, 86, 119, 125
Box Spring Flats, 74
Brinkerhoff, Sidney, 101
Bruderlin, Henry H., 103-104
Burns, "Doc," 93

C

Cabeza de Vaca, Nuñez, 9
California, 20, 36
Carson City Mint, 122
Castañeda, 9, 12
Chalcedony, 74
Christianity, 19
Cibola, Seven Cities of, 9, 10, 11
Clapp, Jay, 92, 93
Clements, William F., 84
Conquistadores, 19
Copper, 73, 75
Coronado, Francisco Vasquez de, 11, 12, 20, 21
Coronado's Children, 11
Cortez, Hernando, 9
Cravey, James A., 84-86, 90
Crawford, Charles, 95, 106, 107
Crook, George, 48, 69
Crooked Mountain, 15

D

Dacite, 15, 73, 74
Deming, 69
Denver Mint, 122
Denver Post, 103
Dobie, J. Frank, 11
Doctor Thorne Mine, 61
Dons' Camp, 124
Dorantes, Andres, 9
Duppa, Darrell, 29
Dutchman, The (see Waltz)
Dutchman's Trail, 125

E

Early, Lynn, 84, 85
Earthquake, 68-71
Edwards, Frank, 97
El Paso, 69
Ely, Sims, 101, 107
Esteban, 9, 10, 11

F

Feldspar, 15
Fernandez, Stanley, 89-90
Ferreira, Benjamin, 89-90
Fifth U.S. Cavalry, 48
First Water, 17, 92, 125
First Water Ranch, 86
Fish Creek, 74, 99
Flagg Foundation, 110, 111
Florence, 25, 64, 69, 91, 95, 121
Florence Blade, 64
Florence Junction, 16, 17, 110
Flores, Bernardo, 89
Florida, 9
Ford, Glenn, 51
Fort McDowell, 61, 68-70
Four Peaks, 95, 96, 107
Fremont Saddle, 76, 108
Frink, Guy "Hematite," 93

G

General Grant claim, 40
General Services Adm., 122
Geologic Map Pinal County, 73
Geology, 73-75
Germans, 25
Geronimo Head Mountain, 74
Gila County, 74
Globe, 15, 16, 103
Gold, 19, 75, 95-107, 121-123
Goldfield, 15, 73, 75
Goodwin, John N., 38, 39
Grajalva, Selso, 27, 28
Gran Quivira, 11, 12
Guadalupe-Hidalgo Treaty, 61
Gulf Coast, 9
Gulf of California, 11

(Index continued on page 142)

Index / 141

H

Hackberry Mesa, 74
Harrier, Franz, 91
Harshbarger, Charles, 93
Heidelberg University, 37-38
Hieroglyphic Canyon, 16, 73, 125
Hieroglyphic Canyon (photo), 30
Hill 3113 (photo), 90
Hohokam, 48
Hooker, Hi, 95, 96
Hopi Indians, 11
Hornblende, 15, 73, 74
Hrdlicka, Ales, 86
Hurkos, Peter, 97, 98

I-J

Inca Indians, 9
Indians, 47-52
International News Service, 104
Jacob, Robert "Crazy Jake," 104, 106, 107
Jerome, 74
Jesuits, 19, 21, 54, 106-107, 114
JF Ranch, 75
JF Trail, 125
Jones, Celeste Maria, 13, 14, 18, 23, 31, 32, 35, 41, 46, 52-54, 66-67, 72, 76-77, 81-82, 91, 92, 94, 108, 118-120, 127

K-L

Kansas, 12
Kelly, Joseph, 88
Kentworthy, Charles, 104, 106
King, Henry, 29
King's Ranch, 16
La Barge Canyon, 106, 107
Landmarks (map), inside front cover
Lee, Jake, 98
Los Angeles, 36, 37, 54
Los Angeles Times, 69
Lead, 75
Limonite, 73, 74
Lost Dutchman Exploration Co., 101, 102
Lost Dutchman Mine, 7, 8, 13, 15, 17, 21, 24, 49, 51, 52, 55-65, 70, 75-80, 83-93, 95-107, 110-117, 124
Lost Dutchman Mining Corp, 100-101

Lost Mines, 78-80
Lost Railroader's Mine, 80
Louie, 13-14, 18, 23, 31-35, 41, 53, 72, 76-77, 82, 108-109, 118-119

M

Macholz, Oscar, 25
Magill, Glenn, 87, 95, 101, 102
Malakoff, 25
Maldonado, Castillo de, 9
Malm, 62
Manganese, 73
Maricopa County, 25, 37, 40, 74, 83, 86, 92
Maricopa County Courthouse, 43
Maricopa Indians, 48-52
Martz, Jack, 92
Massacre gold, 62, 63
Massey, Charles, 87
McFadden, J. R., 86
McNulty, Marjorie, 88
Mendoza, Antonio de, 10, 11
Mesa, 17, 41, 110, 119
Mesa Museum, 110
Metals, 75
Mexicans, 26, 27
Mexico, 9, 19, 20, 21, 36, 61, 63
Mexico City, 9, 10, 11, 12
Middaugh, Laura B., 83, 84
Miami, 15, 16, 103
Mineral investigation, 74
Mining in Arizona, 78
Missionaries, 19
Mohave Indians, 49
Montezuma, 21, 41, 51, 52, 54
Monzonite, 74
Mountain of Foam, 15
Mowry, Walter J., 91-92
Mullins, Jess, 103, 104

N-O

Nadine, 18
National Archives, 122
Navajo, 48
New Mexico, 19
New Spain, 10, 20
New York, 36
Newsweek, 90-91
Niza, Marcos de, 10, 11
Nogales, 21
Oklahoma Securities Com., 102
Ortega, Pedro, 27, 28
Otsuki, James K., 122

P

Palmer Mine, 75
Palomino Mountain, 74
Papago Indians, 20
Parker Pass, 74
Peralta Canyon, 73, 75
Peralta, Don Miguel, 63
Peralta family, 61-62, 85, 112, 114
Peralta Mines, 65
Peralta Stone Maps, 104, 106, 110-117
Peralta Trail, 76, 108, 119, 125
Perlite, 73
Perrin, R. F., 84
Peru, 9
Petrasch, Hermann, 57, 60
Petrasch, Reiney, 57, 60
Petroglyphs, 12
Phoenix, 8, 15, 16, 18, 24, 37, 38, 41, 43, 69, 108, 110, 121
Phoenix Chamber of Commerce, 88, 96
Phoenix Gazette, 51, 55, 106
Phoenix Herald, 26, 28, 29, 30
Phoenix Republican, 52
Picketpost Mountain, 68, 69
Pima Indians, 17, 20, 29, 49, 50, 51, 52, 64
Pima-Maricopa, 49, 50
Pinal County, 15, 74, 83, 86, 92, 102
Piper, Ed, 66-67, 90-91, 118-119, 127
Pizarro, 9
Polzer, Charles W., 110-117
Prescott, 25, 38, 40, 63, 69, 121
Prospect, The, 63
Pyrite, 7

Q-R

Quartz, 73, 74
Quivira, 11, 12
Ray, 103
Raymond, 7-8, 13-14, 18, 23, 41, 54, 72, 76-77, 108, 119-120
Reavis Trail, 125
Reno Mountains, 26
Rhyolite, 15, 73, 74
Rio Grande, 10
Robinson, Will R., 51
Roger's Trough, 125
Roosevelt Lake, 16
Russell, W. H., 92
Ruth, Adolph, 85-87, 90, 102, 107
Ruth, Erwin, 87, 102

(Index continued)

S

Sacramento, 36
Sacramento Mint, 121-122
Saguaro Lake, 16
St. Louis, 36
St. Marie, Robert, 90-92, 127
Salado, 48
Salt River, 16, 26, 96, 98
Salt River Canyon, 48
San Francisco, 36
San Francisco Mint, 122, 123
San Pedro Valley, 11
Santa Fe, 106
Saturday Evening Review, 60-61
Scarborough, Ed, 29
Schroeder, Albert, 47
Scottsdale, 53, 54
Scuelebtz, Meyer, 89
Sierra de la Espuma, 15
Sikorsky, Robert (photo), 8
Silver, 75
Silver King Mine, 15, 16, 28, 93
Silverlock, 62
Simpson, Irene, 123
Skull Cave, 49
Smeitt, C., 25
Sombrero Butte, 63, 87
Sonora, 63
Southwestern Mission Research Center, 110
Spain, 9, 19
Spaniards, 19, 20, 21, 52, 114
Spanish colonial period, 110
Star, Jake, 29
Starar, Andrew (Andy), 42-45, 57
Starar, Jacob, 44
Strange People, 97
Sunset Trail Ranch, 84
Superior, 7, 14, 15, 16
Superstition Allotment, 124
Superstition Mountains, 7-8, 13-18, 23, 25-28, 30-33, 38, 40-41, 43, 45, 47-49, 51-53, 55, 60, 61, 63, 66, 68-71, 73-80, 83-93, 95, 102, 103, 109, 118, 120-127
Superstition Wilderness, 74, 75, 124-126
Superstition Wilderness (map), 6
Symes, J. Foster, 98

T

Tayopa, 21
Teason, Norman, 91
Tempe, 7, 18, 28, 29, 119
Tenochtitlán, 9
Texas, 9
Thomas, Emil W., 56, 58-59
Thomas, Julia, 13, 29, 30, 54, 55-61, 65, 70
Thorne, Abraham, 61, 62, 65
Thorne Mine, 61
Thunder God, 51
Thunder God's Gold, 51
Tonto National Forest, 106, 124
Tonto National Monument, 16
Tortilla Trail, 125
Trails (map), inside front cover
Tucson, 21, 24, 63, 69
Tucson Daily Citizen, 92
Tuff, 73, 74
Twain, Mark, 102
Tweed, H., 29

U-V

U.S. Bureau of Mines, 73-75
U.S. Census, 37
U.S. Forest Service, 72, 124-125
U.S. Mint, 98
U.S. Treasury, 122
Valley of the Sun, 16
Vekol Mine, 64
Vulture Mine, 39

W

Walker, John D., 64, 65
Walnut Grove Mining District, 40
Waltz, Jacob, 13, 17, 21, 24, 25, 27-30, 36-40, 42-45, 51, 52, 54, 55-65, 70, 90, 95, 98, 103, 107, 121 123
Ward, Huse, 61
Weaver's Needle, 15, 17, 22 (photos), 23, 31-35, 41, 46, 52, 62-64, 66-68, 70-75, 81-82, 84, 87, 90-92, 94, 96, 97, 104, 107, 117, 124-127
Wedgelich, John, 25
Wee-Veak-Ah, 49
Weedin, Tom, 64
Weiser, Jacob, 25, 63-65, 70
Wells Fargo Bank, 123
Wells Fargo Stage, 123
When The Red Gods Made Man, 51
Wickenburg, 39
Wickenburg, Henry, 39
Wiggins, Thomas, 102
Wilderness Act, 124
Wilderness area (map), 6
Woolf, Charlie, 95, 107

Y-Z

Yavapai County, 37, 40
Yavapai Indians, 47
Yuma, 69
Zinc, 75
Zuni Indians, 11
Zywotko, Martin G., 87

Order these titles from your book dealer

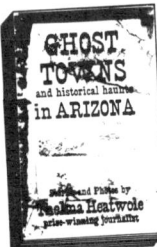

Visit the silver cities of Arizona's golden past. Come along to the towns whose heydays were once wild and wicked. See crumbling adobe walls, old mines, cemeteries, cabins and castles of Arizona's yesteryear. *Ghost Towns in Arizona (by Thelma Heatwole) 144 pages . . . $4.50*

Get acquainted with Arizona's sizzling Indian fry bread, sourdough biscuits, Navajo cake, orange marmalade, Papago beans, beef jerky, prickly pear jelly, cactus candy, chuckwagon steaks, refried beans, salsa, burritos, and much more! *Arizona Cook Book (by Al and Mildred Fischer) 144 pages . . . $3.50*

Take the back roads to Arizona's natural wonders—exotic valleys, natural bridges, idyllic spots, majestic mountains, impassible streams, boxed-in canyons, sparkling lakes—it's all there in *Arizona—Off the beaten path! (by Thelma Heatwole) 144 pages . . . $4.50*

Now, you can prepare these favorite recipes—tacos, tamales, menudo, enchiladas, burros, salsas, frijoles, huevos, almendrado. Home style! Delicious! *Mexican Family Favorites Cook Book (by Maria Teresa Bermudez) 144 pages . . . $5.00*

Follow the daring deeds and exploits of the Earp brothers, Buckey O'Neill, the Rough Riders, Arizona Rangers, cowboys and cattlemen, politicians and miners, shootouts, notorious Tom Horn, Pleasant Valley wars, the "First" American revolution—action-packed true tales of early Arizona! *Arizona Adventure (by Marshall Trimble) 160 pages . . . $5.00*

The lost hopes, the lost lives—the lost gold! Facts, myths and legends of the Lost Dutchman Gold Mine and the Superstition Mountains. Told by a geologist who was there! *Fools' Gold (by Robert Sikorsky) 144 pages . . . $5.00*

Order from your book dealer or direct from publisher.

ORDER BLANK

Golden West Publishers

4113 N. Longview Ave.
Phoenix, AZ 85014

Please ship the following books:

___ Ghost Towns in Arizona ($4.50)
___ Arizona Cook Book ($3.50)
___ Arizona Adventure ($5.00)
___ Arizona—off the beaten path ($4.50)
___ Fools' Gold ($5.00)
___ Mexican Cook Book ($5.00)

I enclose $_____ (Please add $1 per order postage, handling)

Name _____

Address _____

City _____ State ____ Zip ____

This order blank may be photo copied